Environmental Factors Shaping the Soil Microbiome

Environmental Factors Shaping the Soil Microbiome

Editors

Tongmin Sa
Rangasamy Anandham

MDPI • Basel • Beijing • Wuhan • Barcelona • Belgrade • Manchester • Tokyo • Cluj • Tianjin

Editors
Tongmin Sa
Chungbuk National University
Korea

Rangasamy Anandham
Tamil Nadu Agricultural
University
India

Editorial Office
MDPI
St. Alban-Anlage 66
4052 Basel, Switzerland

This is a reprint of articles from the Special Issue published online in the open access journal *Applied Sciences* (ISSN 2076-3417) (available at: https://www.mdpi.com/journal/applsci/special_issues/soil_microbiome).

For citation purposes, cite each article independently as indicated on the article page online and as indicated below:

LastName, A.A.; LastName, B.B.; LastName, C.C. Article Title. *Journal Name* **Year**, *Volume Number*, Page Range.

ISBN 978-3-0365-2510-5 (Hbk)
ISBN 978-3-0365-2511-2 (PDF)

Contents

About the Editors

Tongmin Sa is a Professor of Agricultural Chemistry at Chungbuk National University, South Korea and a Full Member of the Korean Academy of Science & Technology (KAST). He has served as the President of The Korean Society of Soil Science & Fertilizer (KSSSF), the Vice President of the International Scientific Center for Fertilizer (CIEC, Asia), and the Dean of College of Agriculture, Life and Environmental Sciences, Chungbuk National University. He has written 26 books and book chapters and published over 360 articles in prestigious journals, including a recent publication in Nature Reviews Microbiology. He has supervised over 95 graduate students and postdoctoral researchers from many countries. He had attained his undergraduate B.S. and post-graduate M.S. degrees from Seoul National University. He obtained his doctoral degree major in Plant Physiology/Microbiology from the Department of Soil Science, North Carolina State University. His expertise lies in investigations of the mechanisms of microbe-mediated plant growth promotion, plant stress alleviation, microbial diversity, and biofertilizer development. His ongoing research interests include the mechanisms underlying the interaction between ACC deaminase producing endophytic bacteria and plant ethylene emission to enhance tolerance to abiotic stresses. His laboratory is also investigating the structural and functional diversity of bacteria and archaea involved in methane and nitrogen cycling in rice paddies, as influenced by different nutrient regimes, especially in the long-term application of organic and chemical fertilizers. His research group is also working on the development of potential bioinoculants for the alleviation of heat and UV-B radiation stresses on plants. He has also collaborated with other well-known international scientists to study the microbial community profiles associated with major food crops grown worldwide. This will enable the global scientific community to understand the effect of the eco-geographical distribution of plant-microbiome, and to engineer efficient bioinoculants for better crop productivity.

Rangasamy Anandham is an Assistant Professor of Agricultural Microbiology at Tamil Nadu Agricultural University, India, and a life Member of the Association of Microbiologist of India. He obtained a doctoral degree from the Department of Agricultural Chemistry, Soil Microbiology and Fertility Laboratory, Chungbuk National University, Republic of Korea, and Post Doctor from Organic Agricultural Division, National Academy of Agricultural Sciences, Rural Development Administration, Republic of Korea. His expertise lies in the investigation of the tripartite interactions of microbe-insect-mediated plant growth promotion, plant stress alleviation, microbial diversity, molecular bacterial taxonomy, sulfur lithotrophy, and biofertilizer development. His ongoing research interests include gamma irradiation of Bacillus and Streptomyces for overproduction of antimicrobial agents to control plant diseases. He is also investigating the rhizobial and nor rhizobial endophytes for plant growth promotion and alleviation of biotic and abiotic stresses. He has published 75 articles in peer-reviewed journals and authored nearly 15 book chapters. He is an associate editor of The Journal of Microbiology, Review Editor in Frontiers in Microbiology and Madras Agricultural Student Union, which is a 100-year-old journal published from Tamil Nadu Agricultural University, India.

applied
sciences

Editorial

Editorial for Special Issue "Environmental Factors Shaping the Soil Microbiome"

Rangasamy Anandham [1] and Tongmin Sa [2,3,*]

1   Department of Agricultural Microbiology, Tamil Nadu Agricultural University,
    Coimbatore 641003, Tamil Nadu, India; anandhamranga@gmail.com
2   Department of Environmental & Biological Chemistry, College of Agriculture, Life & Environmental Sciences,
    Chungbuk National University, Cheongju 28644, Korea
3   The Korean Academy of Science and Technology, Seongnam 13630, Korea
*   Correspondence: tomsa@chungbuk.ac.kr

Abstract: Soil is a complex system consisting of various abiotic and biotic factors interacting among themselves in a particular time period. These biotic factors are particularly affected by a large number of disturbances or perturbations occurring in the micro-niches. Soil microbiome is the paramount biotic factor responsible for nutrient cycling that in turn determines soil health and quality. However, there are limitations in studying soil systems as there are a number of unknown boxes that need to be checked before understanding their full-fledged contribution to the environment. The microbial diversity in the soil can be affected by salinity, contaminant, fertilization, nutrient accumulation, and cultivation practices. Additionally, plants can also benefit from these changes in microbial community composition, and novel microbial isolates can be used for enhancing their growth under various stress conditions. Collectively, this Special Issue includes various studies that determine key abiotic and biotic factors that can result in changes in microbial community composition in the soil system. These studies are carried out in specified environmental niches rather than microcosms, which provides a broader context in which to understand microbial dynamics.

Keywords: soil microbial community; environmental factors; long-term fertilization; soil contamination; microbial diversity; culture-independent analysis

Citation: Anandham, R.; Sa, T. Editorial for Special Issue "Environmental Factors Shaping the Soil Microbiome". *Appl. Sci.* **2021**, *11*, 10387. https://doi.org/10.3390/app112110387

Received: 26 October 2021
Accepted: 2 November 2021
Published: 5 November 2021

Publisher's Note: MDPI stays neutral with regard to jurisdictional claims in published maps and institutional affiliations.

The soil microbiome comprises one of the most important and complex components of all terrestrial ecosystems as it harbors millions of microbes, including bacteria, fungi, archaea, viruses, and protozoa. Together, these microbes and environmental factors contribute to shaping the soil microbiome, both spatially and temporally. This Special Issue covers the nutrient amendments, application of compost, discharge of azo dye effluents, mining, soil salinity, soil acidity, and plant genotypes that affect the soil microbial community structures. Additionally, it provides a glimpse into dominant microbial communities from these environments that may be tapped for biotechnological applications.

Microbial populations in soil are determined by various factors such as soil depth, organic matter, porosity, oxygen and carbon dioxide concentration, soil pH, etc. Factors that influence microorganisms' roles in nutrient building and cycling in soil and organic matter decomposition are of unique interest. These microbial entities have the potential to be utilized in various biotechnological applications that are necessary to improve soil health and quality. Microorganisms decompose organic matter, and detoxifying toxic substances, fixing nitrogen, and transforming nitrogen, phosphorous, potassium, and other secondary and micronutrients are the major biochemical activities performed by microbes in the soil. They play a major role in supporting life forms in micro-niches by supplying essential nutrients and carrying out bio-transformations that are essential for the survival of living organisms. Soil fertility is considered an important factor for better agricultural production. Microbial community composition and diversity of agricultural soils primarily depend on management practices. The application of compost to agricultural fields is known to increase soil fertility, which can also help to enhance agricultural productivity. The relative abundance of a few groups of bacteria, such as Firmicutes, Actinobacteria, and Proteobacteria, is significantly higher in compost-amended soil, and several of these bacteria are reported to be beneficial. Thus, a combined application of compost and inorganic fertilizers may be a good way to keep up with agricultural productivity while keeping the environmental balance. The shift in bacterial community composition through compost amendment especially leads to an increase in decomposition-related enzymes, which are a primary factor in enhancing nutrient accumulation, leading to fertile soil that supports better plant growth. Additionally, compost amendment has also been reported to increase the abundance of enzymes that are important during various developmental stages of plants [1].

Soil properties are one of the major factors determining the growth of vegetation. The natural habitat is used in various parts of the world for the cultivation of plants that are important for manufacturing traditional medicines. The soil's properties drive the selection of the dominant bacterial community profiles, which eventually determines the soil quality and fertility, making it conducive to supporting vegetation. The abundance of a preferential bacterial community assists in better productivity of particular types of vegetation [2]. Rhizosphere bacterial diversity is known to affect plant health, and communities with a higher diversity are generally better able to withstand invasion by pathogens and possess higher amounts of plant-growth-promoting bacteria (PGPB). The preferential biodiversity often relates to increasing levels of ecosystem functioning and services. Soil communities are also affected by plant characteristics, primarily through the production of root exudates. Bacterial communities are also known to differ according to plant genotypes and hosts. This was supported by the findings of Vink et al. [3], who concluded that grape cultivars are able to recruit (i.e., attract and select) different, potentially beneficial genera. Further, they summarized that *Desertibacter* and *Rhodothalassium* occurred in relatively high abundance in vine cultivars Calandro and Villaris. Only one genus (*Stenotrophomonas*) was in high abundance in cultivars Felicia and Reberger. However, this effect is by no means ubiquitous, and plant genotypic differences do not always lead to significant differences in microbiomes in the rhizosphere. Even so, this puts forward a novel idea about the preferential plant–microbe interaction based on plant genotypes. These preferential microbial associations can help in identifying key bacterial taxa in the particular environmental niche and determine the particular plant genotype that can be cultivated in that region for better productivity.

Exogenous nutritional inputs are an inevitable process in crop production and can change the structures of soil bacterial communities. Among the several nutrients, nitrogen and phosphorous are the most limiting nutrients in agricultural productivity. This Special Issue cautions that higher deposition or accumulation of N in the urban green space deeply affects the patterns of bacterial diversity and community structure. Nitrogen addition has a negative impact on bacterial richness and diversity in urban green space. The decrease in biodiversity induced by N deposition may pose a serious threat to the stability of urban soil ecosystems, which emphasizes the necessity of thorough and concerted studies to prompt adequate policies to counteract these globally increasing threats [4]. Phosphorus removal from phosphorus-enriched soils (PES), such as agricultural lands, grasslands, and phosphate mining regions, is a major cause of eutrophication in aquatic environments and an increasing environmental problem worldwide. Phytoremediation with *Erianthus rufipilus*, *Coriaria nepalensis*, and *Pinus yunnanensis* is one of the most promising technologies for the removal of excess P in agricultural systems. In a study, rhizospheric microbial communities are shaped not by the plant species but by soil water content, soil organic matter, and total nitrogen contents of the P-contaminated sites [5]. This eventually reveals the complex interaction between different niches and ecosystems in determining the soil microbial diversity.

Discharge of untreated wastewater is one of the major problems in various countries. The use of azo dyes in textile industries is one of the key xenobiotic compounds that affect both soil and water ecosystems and result in a drastic effect on the microbial communities. Orathupalayam dam, which is constructed over the Noyyal river in Tamil Nadu, India, has become a sink of wastewater from the nearby textile industries, and this polluted site especially supports the abundance of *Saccharibacteria*; hence, enrichment isolation of this bacterium could be used to degrade the azo dyes and remediation of textile effluent degrading sites [6]. The agricultural soils that use the dam water for irrigation purposes have also shown a minor shift in bacterial diversity that implies potential contamination due to azo dye compounds. However, determining the key bacterial taxa from these contaminated soils can assist in designing potential bioremediation techniques that are sustainable and environmentally friendly. Further studies designing potential biotechnological technologies will not be possible if the microbiomes of the particular environment are not scrutinized beforehand. Hence, studying these particular sites is of utmost importance for putting forward industrially important products.

Mine heaps and mine wastes created by the mining industry are some of the extreme habitats made by anthropogenic activity. Interestingly, mine heaps create an environment with specific ecological conditions for plant adaptation. They are characterized by the lack of soil, nutrients, and moisture, as well as an absence of a humus layer. Orchids represent a unique group of plants that are well adapted to these extreme conditions. In this Special Issue, Böhmer et al. [7] addressed the microbial diversity of orchids in polluted sites. They found that the pH of the initial soils does not significantly affect the presence of fungi and bacteria. Similarly, toxic elements (e.g., As, Pb, Cr, Ni, Co, Cu, Fe) do not affect the occurrence of fungi, bacteria, or orchids. Moreover, microbial communities also provide a huge benefit for orchids to be able to grow in these polluted areas. Additionally, it can be concluded that some of these microbial communities possess huge biotechnological potential in bioremediation of heavy-metal-contaminated areas. This addresses another key feature of microbial community composition in remediating toxic compounds that are a result of anthropogenic activity. Heavy metals have been a matter of concern in various mining-rich locations that possesses a high threat to human health. The chemical remediation of heavy metals is expensive, and exploring an unorthodox pathway can prove to be a sustainable approach.

Soil salinity is a severe agronomical, ecological, and socioeconomic concern in most arid and semi-arid regions of the world. It is estimated that salinization will threaten more than 50% of arable land worldwide by 2050. Hence, this silent hazard will continue to threaten agricultural sustainability, food security, ecosystem stability, human health, and income generation. The use of plant-growth-promoting bacteria for salt stress alleviation is practiced widely. Since it has no harmful impact on the environment, this method has huge benefits and is widely applied. Arbuscular Mycorrhizal Fungi (*Glomus mosseae*) inoculation, alone or in combination with plant-growth-promoting bacteria (*Bacillus amyloliquefaciens*), increased biomass accumulation, morphological characteristics, photosynthetic capacity, and rhizospheric soil enzyme activities in saline soils [8]. Acidic soils occupy around 40% of the total agricultural lands worldwide, representing one of the most important limiting factors of agriculture production. Both liming and plant residue incorporation are widely used practices for the amelioration of acidic soils—however, the difference in their effects is still not fully understood, especially regarding the microbial community. In this Special Issue, Li et al. [9] demonstrated that liming was effective in elevating soil pH, while plant residue incorporation exerted a comprehensive influence not only on soil pH but also on soil enzyme activity and microbial community. Nannipieri [10] critically reviewed the soil as a biological system. Still knowledge gaps can be found. This Special Issue shows that technology-driven and hypothesis-driven research should be combined in order to fill the remaining gaps. Particularly imaginative research should address the simulation of the soil microenvironment so as to understand which factors regulate microbial activities in micro-niches.

Author Contributions: Writing—original draft preparation, T.S., R.A.; writing—review and editing, T.S. All authors have read and agreed to the published version of the manuscript.

Funding: This work did not receive any external funding.

Institutional Review Board Statement: Not applicable.

Informed Consent Statement: Not applicable.

Data Availability Statement: Not applicable.

Conflicts of Interest: The authors declare no conflict of interest.

References

1. Kim, S.; Samaddar, S.; Chatterjee, P.; Roy Choudhury, A.; Choi, J.; Choi, J.; Sa, T. Structural and Functional Shift in Soil Bacterial Community in Response to Long-Term Compost Amendment in Paddy Field. *Appl. Sci.* **2021**, *11*, 2183. [CrossRef]
2. Kim, K.; Kim, H.J.; Jeong, D.H.; Huh, J.H.; Jeon, K.S.; Um, Y. Correlation between Soil Bacterial Community Structure and Soil Properties in Cultivation Sites of 13-Year-Old Wild-Simulated Ginseng (*Panax ginseng* C.A. Meyer). *Appl. Sci.* **2021**, *11*, 937. [CrossRef]
3. Vink, S.N.; Dini-Andreote, F.; Höfle, R.; Kicherer, A.; Salles, J.F. Interactive Effects of Scion and Rootstock Genotypes on the Root Microbiome of Grapevines (*Vitis* spp. L.). *Appl. Sci.* **2021**, *11*, 1615. [CrossRef]
4. Mo, L.; Zanella, A.; Chen, X.; Peng, B.; Lin, J.; Su, J.; Luo, X.; Xu, G.; Squartini, A. Effects of Simulated Nitrogen Deposition on the Bacterial Community of Urban Green Spaces. *Appl. Sci.* **2021**, *11*, 918. [CrossRef]
5. Hu, Y.; Duan, C.; Fu, D.; Wu, X.; Yan, K.; Fernando, E.; Karunarathna, S.C.; Promputtha, I.; Mortimer, P.E.; Xu, J. Structure of Bacterial Communities in Phosphorus-Enriched Rhizosphere Soils. *Appl. Sci.* **2020**, *10*, 6387. [CrossRef]
6. Krishnamoorthy, R.; Roy Choudhury, A.; Arul Jose, P.; Suganya, K.; Senthilkumar, M.; Prabhakaran, J.; Gopal, N.O.; Choi, J.; Kim, K.; Anandham, R.; et al. Long-Term Exposure to Azo Dyes from Textile Wastewater Causes the Abundance of *Saccharibacteria* Population. *Appl. Sci.* **2021**, *11*, 379. [CrossRef]
7. Böhmer, M.; Ozdín, D.; Račko, M.; Lichvár, M.; Budiš, J.; Szemes, T. Identification of Bacterial and Fungal Communities in the Roots of Orchids and Surrounding Soil in Heavy Metal Contaminated Area of Mining Heaps. *Appl. Sci.* **2020**, *10*, 7367. [CrossRef]
8. Pan, J.; Huang, C.; Peng, F.; Zhang, W.; Luo, J.; Ma, S.; Xue, X. Effect of Arbuscular Mycorrhizal Fungi (AMF) and Plant Growth-Promoting Bacteria (PGPR) Inoculations on *Elaeagnus angustifolia* L. in Saline Soil. *Appl. Sci.* **2020**, *10*, 945. [CrossRef]
9. Liu, X.; Feng, Z.; Zhou, Y.; Zhu, H.; Yao, Q. Plant Species-Dependent Effects of Liming and Plant Residue Incorporation on Soil Bacterial Community and Activity in an Acidic Orchard Soil. *Appl. Sci.* **2020**, *10*, 5681. [CrossRef]
10. Nannipieri, P. Soil Is Still an Unknown Biological System. *Appl. Sci.* **2020**, *10*, 3717. [CrossRef]

Communication

Structural and Functional Shift in Soil Bacterial Community in Response to Long-Term Compost Amendment in Paddy Field

Sookjin Kim [1,2,†], Sandipan Samaddar [1,†,‡], Poulami Chatterjee [1,§], Aritra Roy Choudhury [1], Jeongyun Choi [1], Jongseo Choi [3] and Tongmin Sa [1,*]

[1] Department of Environmental and Biological Chemistry, Chungbuk National University, Cheongju 28644, Korea; ksj8827@korea.kr (S.K.); ssamaddar@ucdavis.edu (S.S.); pchatterjee@ucdavis.edu (P.C.); aritraroy@chungbuk.ac.kr (A.R.C.); jychoi@chungbuk.ac.kr (J.C.)
[2] Planning and Coordination Division, National Institute of Crop Science, RDA, Wanju 55365, Korea
[3] Crop Cultivation and Environment Research Division, National Institute of Crop Science, RDA, Suwon 16429, Korea; hbell7@korea.kr
* Correspondence: tomsa@chungbuk.ac.kr
† These authors have contributed equally to this work.
‡ Present address: Department of Land, Air and Water Resources, University of California, Davis, CA 95616, USA.
§ Present address: Department of Microbiology and Molecular Genetics, University of California, Davis, CA 95616, USA.

Citation: Kim, S.; Samaddar, S.; Chatterjee, P.; Roy Choudhury, A.; Choi, J.; Choi, J.; Sa, T. Structural and Functional Shift in Soil Bacterial Community in Response to Long-Term Compost Amendment in Paddy Field. *Appl. Sci.* **2021**, *11*, 2183. https://doi.org/10.3390/app1105 2183

Academic Editor: Stefano Castiglione

Received: 18 December 2020
Accepted: 4 February 2021
Published: 2 March 2021

Abstract: Microbial community composition and diversity of agricultural soils primarily depend on management practices. The application of compost on agricultural fields is known to increase soil fertility, which can also help to enhance agricultural productivity. The effects of long-term application of compost along with nitrogen (N), phosphorus (P), and potassium (K) (+Compost) on soil bacterial diversity and community profiles were assessed by amplicon sequencing targeting the 16S rRNA gene of bacteria and compared with those on soils that received only NPK but not compost (−Compost). Ordination plot showed treatments to cluster differently, implying changes in community composition, which were validated with taxonomical data showing Firmicutes, Actinobacteria, and their related classes to be significantly higher in +Compost than in −Compost soils. The predicted abundance of functional genes related to plant growth promotion, development, and decomposition was significantly higher in compost-amended soil than in soils without compost. The results are of particular importance as they provide insights into designing management practices to promote agricultural sustainability.

Keywords: long-term fertilization; next-generation sequencing; bacterial diversity; plant growth

1. Introduction

The diverse and abundant population of soil bacteria plays a major role in the functioning of the ecosystem; however, different agricultural management strategies primarily drive the bacterial community composition and functioning [1,2]. Nutrient amendments in agricultural soils are usually used to improve plant productivity, but overuse of fertilizers, like inorganic N, can affect the soil quality by deteriorating soil fertility [3], decreasing crop yield [4], and affecting bacterial diversity [5,6].

With the increasing importance of soil microbes, which include bacteria, in maintaining soil quality [7], understanding soil microbial processes under different management schemes is recognized to be important for the sustainability of agricultural ecosystems. Soil fertility management based on the use of organic fertilizers can promote microbial processes and increase crop yield [8]. Increment in microbial processes can help biogeochemical cycles and nutrient cycling [9], which can assist in the enhancement of crop productivity [10]. Extensive research has indicated the beneficial effect of organic matter application in enhancing bacterial diversity and positive interaction with plants [11,12].

Livestock manure and crop residue are two major types of organic matter that have the potential to mitigate soil degradation caused by long-term chemical fertilization. Several studies have addressed the impacts of livestock manure and plant residues on soil-dwelling bacterial communities [13,14]. Crop residues are the most abundant, cheapest, and most readily available organic waste to be biologically transformed, with rice straw being one of them, in many rice-growing countries. They have been used as a common organic material as they contain numerous mineral nutrients, such as nitrogen, phosphorus, potassium, and silicate [15]. The application of straw in soil has been used to improve the activity of soil bacterial community and promote soil nitrogen and carbon sequestration potential [16]. Additionally, livestock manure has shown similar characteristics as crop residues [17]. However, there have been limited studies that concentrate on the effect of compounded compost made of rice straw and livestock manure on soil bacterial diversity and community profiles.

Thus, this study was conducted to compare and characterize the bacterial diversity and community composition of paddy soils treated with NPK combined with rice straw and livestock manure (+Compost) and that with only conventional fertilizers without compost (−Compost). We hypothesized that the additional compost amendment will result in changes in the structural and functional composition of the bacterial community.

2. Materials and Methods

The experimental field (1230 m^2) was located at the Paddy Crop Research Division, Department of Southern Area Crop Science, National Institute of Crop Science, Republic of Korea. The plots within the experiment were managed and fertilized as mentioned previously by Samaddar et al. [18]. Briefly, the fertilization trial was initiated in 1967, and rice (*Oryza sativa* cv. Hwayeong) was cultivated as a single crop. Out of several fertilization treatments, two different treatments with three replicate plots (10 m × 10 m each) for each treatment arranged in a completely randomized manner were used in this study. The treatments were (1) inorganic NPK fertilized soil sample (−Compost) and (2) NPK amended with compost (10 tons·ha^{-1}·year^{-1}) fertilized soil sample (+Compost). NPK fertilizers were applied as urea, fused superphosphate, and potassium chloride at the rate of 120, 80, and 80 kg·ha^{-1} from 1967 to 1972. Since 1973, N, P, and K have been applied at a higher rate of 150, 100, and 100 kg·ha^{-1}, respectively. Compost was prepared by decomposing rice straw and cattle manure for a period of 6 months, and it contained 431, 19.8, 5.2, and 29.1 g·kg^{-1} of total C, N, P, and K, respectively. A total of five individual subsamples (~20 cm depth) were collected from each of the three replicates of individual treatments in April 2016 and were pooled to make a composite sample (~100 g) for each replicate. Thus, six individual samples were collected. Soil samples were sieved (<2 mm) and stored in sterile bags and transported to the laboratory on ice. Soil DNA was extracted using 0.5 g of fresh soil sample by using a PowerSoil DNA Isolation Kit (Mo Bio Laboratories, Carlsbad, CA, USA), and DNA was stored at −20 °C for all downstream analysis. The 16S rRNA copy number was estimated using quantitative PCR (qPCR) with the primers 517F/1028R [19,20] following conditions described in detail by Samaddar et al. [18]. Briefly, qPCR amplification targeting the 16S rRNA gene of bacteria was performed using the primers 517F (5′-GCCAGCAGCCGCGGTAA-3′) and 1028R (5′-CGACARCCATGCASCACCT-3′) with a Rotor-Gene® Q (Qiagen, Foster City, CA, USA) in 10 μL of reaction mixture containing 5 μL of Maxima SYBR Green Master Mix (Thermo Fisher Scientific Inc., Waltham, MA, USA), 0.4 μM of each of forward and reverse primers, 1 μL of template DNA, and 3.6 μL of sterile distilled water. The reaction conditions for the qPCR were as follows: 1 cycle at 95 °C for 10 min, 45 cycles of denaturation at 95 °C for 30 s, annealing at 60 °C for 40 s, and extension at 72 °C for 45 s. All sample and standard curve measurements were performed in triplicate, and negative controls were included in every qPCR run. For standard curve preparation, the 16S rRNA fragment from *Pseudomonas mendocina* PC1 was amplified with the same primer pair under the same thermal conditions. Rotor-Gene Series Software v. 2.0.2 (Qiagen, Foster City, CA, USA) and LinRegPCR program v. 2017.0 (Academic Medical Centre, Amsterdam, Netherlands) [21] were used to analyze the obtained data. The qPCR standard curve efficiency was 1.866 ± 0.031 ($n = 12$). Additionally, the amplification efficiency of each sample was considered for calculating the 16S rRNA gene copy numbers. The isolated DNA was subjected to high-throughput Illumina MiSeq sequencing at ChunLab, South Korea. The V3–V4 regions of 16S rRNA genes were targeted and amplified using the primers 341F (5′-TCGTCGGCAGCGTCAG ATGT GTATAAGAGACAG<u>CCTACGGGNGGCWGCAG</u>-3′; underlined sequences indicating the target) and 805R (5′-GTCTCGTGGGCTCGGAGATGGTATAAGAGACAG <u>GACTACHVGGTATCTAATCC</u>-3′). The raw data obtained were analyzed using a Mothur pipeline v. 1.39.5 [22] similarly as performed in our previous studies [18,23]. The raw sequences were deposited in a Sequence Read Archive (SRA) dataset at NCBI (National Center for Biotechnology information) under accession numbers SRP127951 (−Compost) and SRP298800 (+Compost). Analyzed data were normalized to a minimum number of reads prior to calculation of Shannon and Chao indices in Mothur. Principal coordinate analysis (PCoA) was performed in Mothur. Tukey's test was used wherever necessary to calculate the differences between the means, which were considered significant at $p < 0.05$ using SAS version 9.4 [24]. Bacterial community was characterized using the linear discriminant analysis (LDA) effect size (LEfSe) tool for biomarker discovery (http://huttenhower.sph.harvard.edu/lefse/, accessed on 25 February 2021) [25]. The PICRUSt tool was used to predict functional profiles of bacterial communities from the bacterial 16S rRNA abundance data [26].

3. Results

The bacterial 16S rRNA abundance, as determined by qPCR, increased significantly by nearly threefold in +Compost soil compared with −Compost soil (Figure 1a). Analysis of sequence data yielded on average 782 operational taxonomic units (OTUs) for −Compost and 750 OTUs for +Compost, which were not significantly different. Additionally, no significant differences for diversity and richness estimates were observed between the treatments (Table 1). A 98% coverage from Good's coverage estimator and rarefaction curve (Figure S1) showed that sampling was sufficient to estimate those indices. Principal coordinate analysis (PCoA), performed to estimate the effects of studied treatments on bacterial community composition, showed differences in ordination patterns of the treatments with PC1 explaining 28% of variance and PC2 explaining 22% of variance (Figure 1b).

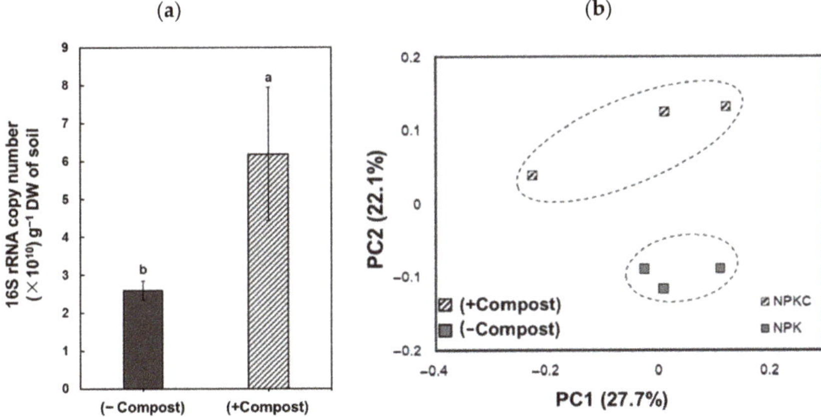

Figure 1. (**a**) The 16S rRNA copy numbers of soil bacterial communities estimated by qPCR. (**b**) The principal coordinate analysis (PCoA) showing the clustering of bacterial communities between soil samples. Values for 16S rRNA data are plotted with mean; error bars indicate standard errors; different letters on plot mean significant difference ($p < 0.05$) between treatments according to Tukey's test.

Table 1. Summary data for 16S rRNA sequencing results and α-diversity of the soil samples.

Treatment	No. of Sequences	Goods Coverage (%)	No. of OTUs	Chao	Shannon's Index
−Compost	6567	98.6	782 ± 6.71 [a]	816 ± 8.2 [a]	6.02 ± 0.02 [a]
+Compost	6567	98.8	750 ± 13.5 [a]	775 ± 15 [a]	5.98 ± 0.01 [a]

[a] Values are not significant at $p < 0.05$ among the treatments according to Tukey's test.

At taxonomic levels, the sequences from the soils were annotated to 56 different phyla, of which the 20 most abundant phyla are represented in Figure 2a. Proteobacteria were the most abundant bacterial phylum in both the studied soil groups, followed by Chloroflexi, Actinobacteria, Verrucomicrobia, Firmicutes, Bacteroidetes, Nitrospirae, uncultured-OD1, Gemmatimonadetes, uncultured-TM7, Cyanobacteria, Chlorobi, Elusimicrobia, Fibrobacteres, Chlamydiae, Spirochaetes, uncultured-WS3, and Armatimonadetes. Of all phyla, the abundance of Chloroflexi and Firmicutes varied significantly with Chloroflexi significantly abundant in −Compost soils while Firmicutes significantly dominating +Compost soils. At the bacterial order level (Figure 2b), Actinomycetales were the most abundant, followed by other groups. However, Rhizobiales and Clostridiales were significantly abundant in +Compost, whereas the abundance of uncultured-SJA-15 increased significantly in −Compost soils. LEfSe analysis (Figure 3) demonstrated that −Compost soils had significantly higher abundance of the phylum Chloroflexi and their related class Anaerolineae and the phylum Acidobacteria and their related order Acidobacteriales compared with −Compost soils. On the other hand, +Compost soils had higher abundance of Actinobacteria along with the family Micrococcaceae and genus *Arthrobacter*, and Firmicutes along with the class Bacilli and order Bacillales. Additionally, abundance of the class Alphaproteobacteria, their related order Rhizobiales, family Hyphomicrobiaceae, and genus *Rhodoplanes* increased significantly in +Compost soils compared with −Compost soils. The predicted abundance of genes encoding enzymes related to plant growth promotion, development, and fatty acid biosynthesis was significantly increased in +Compost soils when compared with −Compost soils (Figure 4).

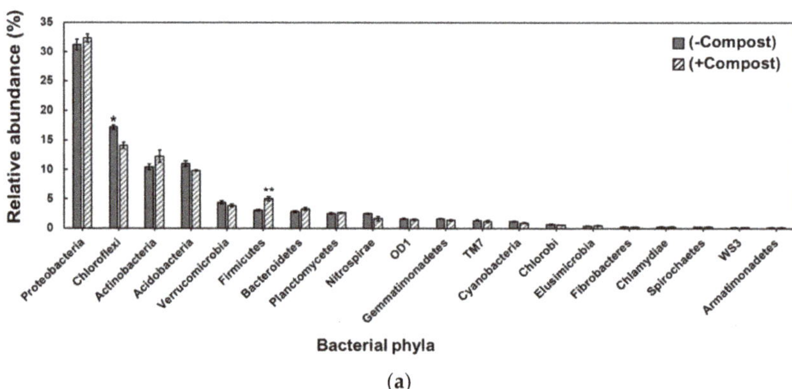

(a)

Figure 2. *Cont.*

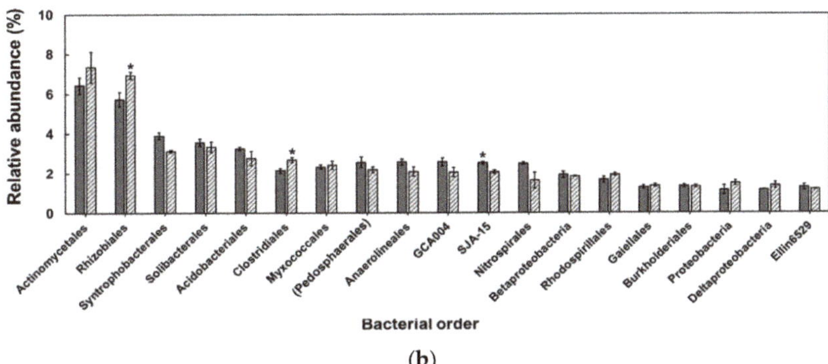

Figure 2. The relative abundances of the 20 most abundant bacterial (**a**) phyla and (**b**) orders in the studied soil samples. The phylum/order significantly different according to ANOVA is marked by asterisk (*: $p < 0.05$; **: $p < 0.01$).

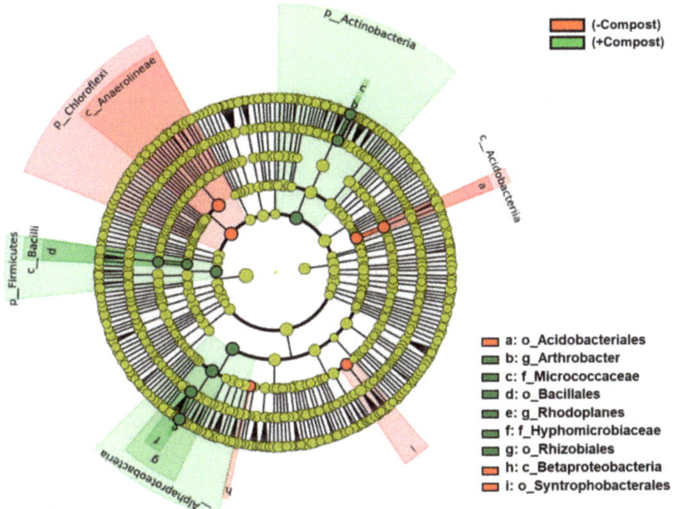

Figure 3. The significantly different taxa of bacteria between two soil samples depicted by linear discriminant analysis effect size (LEfSe) method. The significantly different abundant taxa among treatments are denoted by colored dots, and from the center to the outer circle, they represent the kingdom, phylum, class, order, family, and genus. The trends of the significantly different taxa are represented by colored dots.

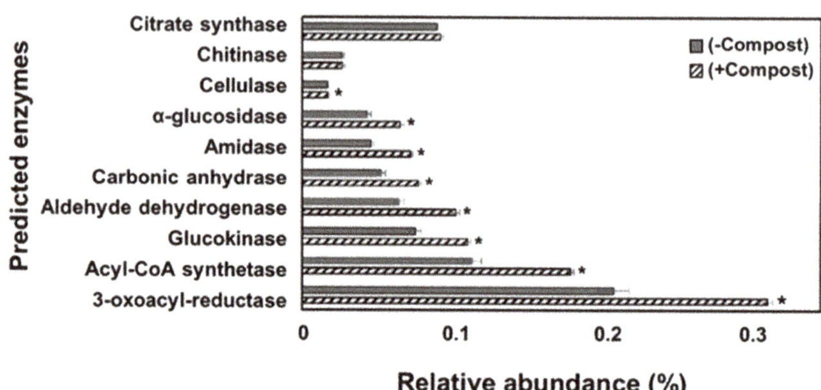

Figure 4. The abundance of genes encoding enzymes in the two soil samples related to plant growth promotion, development, and fatty acid biosynthesis predicted by the PICRUSt. Error bars indicate standard errors; the relative abundance of enzyme-encoding genes with significantly differed between treatments are marked by asterisk (*: $p < 0.05$).

4. Discussion

Investigation of soil microbial communities, considered as indicators of soil quality, [7] showed altered compositional and functional profiles of bacterial community in +Compost and −Compost soils.

The bacterial 16S RNA abundance increased in +Compost soil compared with −Compost soil, which is in agreement with the observations of Li et al. [27], where the 16S rRNA gene abundance increased in soils amended with manure and conventional fertilizer compared with conventional fertilizer alone. Besides, compost amendment did not alter the microbial diversity or richness, but differences in community composition were observed from taxonomical data as also observed in other studies [28,29]. Diversity or richness might not always alter with changes in community composition, as differences in abundance of some taxonomic groups may be compensated by differences in abundance of other taxonomic groups [30]. Additionally, it is true that these estimates are powerful tools and provide us an ecological trend, but it also important to keep in mind that univariate analyses, like alpha diversity estimates, are just a step in the line of scientific query and do not provide a definite answer to community outcomes [31] or composition, which was observed to be altered in the present study.

Proteobacteria were the most abundant phylum belonging to both the studied soil groups [32,33]. However, the abundance of Chloroflexi was observed to be significantly higher in −Compost soils along with their order, the uncultured bacterium SJA-15. Zhalnina et al. [34] recently reported that a decrease in pH and an increase in total nitrogen can contribute to a decrease in Chloroflexi population, as they might be following an oligotrophic lifestyle. Additionally, the amendment of straw and manure results in a decrease in soil pH and an increase in total nitrogen content in soil [35], which might have resulted in a decrease in Chloroflexi population in +Compost soils. The relative abundance of Firmicutes and their order Clostridiales was observed to be higher in +Compost soils, which is important as Firmicutes are regarded as the main phylum that consists of decomposers and are important for the conversion of organic matters [36]. These observations also get support from Sharmin et al. [37], where the abundance of Firmicutes increased in a sugarcane processing plant, which encompasses a large amount of plant organic matters. Actinobacteria and their related class and genus were also significantly abundant in +Compost soils, which draws support from studies where Actinobacteria were sensitive to management strategies [18,38]. A significant increase in the abundance of Rhizobiales in +Compost soil is interesting as they are one of the most important bacterial orders responsible for nitrogen fixation and the enhancement of the total nitrogen content of the soil [39]. Changes in bacterial community composition might also have contributed to changes in the functional profiles of the bacterial community as observed from the predicted abundance of genes encoding enzymes, which were significantly higher in +Compost soil when compared with −Compost soil. One of the limitations of amplicon gene sequencing is that it does not provide information on what the microbes are doing, so the use of the PICRUSt tool [26], which uses an ancestral-state reconstruction algorithm to predict the presence of gene families and then combines them to estimate composite metagenome, is expected to provide an idea about the functional composition of microbes, which is believed to be informative. This tool has been proved to be highly accurate in predicting a community's functional capabilities from 16S abundance profiles [26,40] and has been popular recently [41–44]. The abundance of genes encoding enzymes, which were relatively higher in +Compost compared with −Compost, was mostly related to decomposition and plant growth promotion and is well known to be synthesized by Proteobacteria and Firmicutes [45]. Amidases, which increased significantly in +Compost soil, are characteristic of Actinobacteria [18] and are important for the synthesis of indole-3-acetic acid, an important plant-growth-promoting hormone [46]. Likewise, bacterial citrate synthases, which were relatively higher in abundance in +Compost soil, have shown enhanced plant growth under nutrient-limited soils [47,48]. On the other hand, cellulases, which also increased in +Compost soil, are important for breaking down cellulose into monosaccharides or shorter polysaccharides [49], which improves soil fertility and plant growth through accelerated straw decomposition [50]. The requirement of higher amount of ATP in the decomposition process might have resulted in an increase in the abundance of glucokinase in +Compost soils [51]. On the other hand, compost addition also improved aldehyde dehydrogenase activity, which is known to provide stress tolerance to plants [52]. Furthermore, enzymes known to act on plant development, like acyl-CoA synthetase, which is essential for cuticle development [53], microsporogenesis [54], and pollen development [55], also increased in +Compost soils. Lastly, the increase in the predicted abundance of the fatty acid biosynthesis-specific enzyme 3-oxoacyl reductase, which is known to improve seed yield [56], was also higher in compost-amended soils.

5. Conclusions

Long-term amendment of compost altered the bacterial community composition both structurally and functionally. The relative abundances of a few groups of bacteria, like Firmicutes, Actinobacteria, and Proteobacteria, were significantly higher in compost-amended soil, of which several are reported to be beneficial. Moreover, the predicted abundance of genes coding enzymes related to decomposition, plant growth promotion, and development increased in +Compost soils compared with −Compost soils alone. Thus, a combined application of compost and inorganic fertilizers might be a good way to keep up with the agricultural productivity while keeping the environmental balance.

Supplementary Materials: The following are available online at https://www.mdpi.com/2076-3417/11/5/2183/s1, Figure S1: Rarefaction curves of bacterial 16S rRNA sequences obtained from the studied soil samples.

Author Contributions: S.K. and T.S.: conceptualization of the study and funding acquisition; S.K. and S.S.: design of experiments and performance of experiments; S.K. and S.S.: analysis of sequencing data; P.C. and J.C. (Jongseo Choi): assistance in soil sampling and preliminary experiments; S.K., S.S., and P.C.: writing of the manuscript; A.R.C. and J.C. (Jeongyun Choi): assistance in experiments and writing of the manuscript; T.S.: critical review and editing. All authors have read and agreed to the published version of the manuscript.

Funding: This work was supported by the Research Program for Agriculture Science and Technology Development in Rural Development Administration (Project No. PJ01135101), Republic of Korea.

Institutional Review Board Statement: Not applicable.

Informed Consent Statement: Not applicable.

Data Availability Statement: The data and analyses from the current study are available from the corresponding authors upon reasonable request. The raw reads were deposited in SRA archives and can be accessed by accession nos. SRP127951 (−Compost) and SRP298800 (+Compost).

Acknowledgments: The authors would like to thank Jaak Truu from the University of Tartu, Estonia, for his guidance towards performing and analyzing the qPCR data.

Conflicts of Interest: The authors declare no conflict of interest.

References

1. Fierer, N.; Lauber, C.L.; Ramirez, K.S.; Zaneveld, J.; Bradford, M.A.; Knight, R. Comparative metagenomic, phylogenetic and physiological analyses of soil microbial communities across nitrogen gradients. *ISME J.* **2012**, *6*, 1007–1017. [CrossRef]
2. Zhang, Y.; Dong, S.; Gao, Q.; Liu, S.; Zhou, H.; Ganjurjav, H.; Wang, X. Climate change and human activities altered the diversity and composition of soil microbial community in alpine grasslands of the Qinghai-Tibetan Plateau. *Sci. Total Environ.* **2016**, *562*, 353–363. [CrossRef]
3. Janssens, I.A.; Dieleman, W.; Luyssaert, S.; Subk, J.A.; Reichstein, M.; Ceulemans, R.; Ciais, P.; Dolman, A.J.; Grace, J.; Matteucci, G.; et al. Reduction of forest soil respiration in response to nitrogen deposition. *Nat. Geosci.* **2010**, *3*, 315–322. [CrossRef]
4. Yadav, R.L.; Yadav, D.S.; Singh, R.M.; Kumar, A. Long term effects of inorganic fertilizer inputs on crop productivity in a rice-wheat cropping system. *Nutr. Cycl. Agroecosyst.* **1998**, *51*, 193–200. [CrossRef]
5. Zhou, J.; Guan, D.; Zhou, B.; Zhao, B.; Ma, M.; Qin, J.; Jiang, X.; Chen, S.; Cao, F.; Shen, D.; et al. Influence of 34-years of fertilization on bacterial communities in an intensively cultivated black soil in northeast China. *Soil Biol. Biochem.* **2015**, *90*, 42–51. [CrossRef]
6. Zhou, J.; Jiang, X.; Zhou, B.; Zhao, B.; Ma, M.; Guan, D.; Li, J.; Chen, S.; Cao, F.; Shen, D.; et al. Four years of nitrogen fertilization decreases fungal diversity and alters fungal community composition in black soil in northeast China. *Soil Biol. Biochem.* **2016**, *95*, 135–143. [CrossRef]
7. Liu, M.; Sui, X.; Hu, Y.; Feng, F. Microbial community structure and the relationship with soil carbon and nitrogen in an original Korean pine forest of Changbai Mountain, China. *BMC Microbiol.* **2019**, *19*, 218. [CrossRef]
8. Mäder, P.; Fliessbach, A.; Dubois, D.; Gunst, L.; Fried, P.; Niggli, U. Soil fertility and biodiversity in organic farming. *Science* **2002**, *296*, 1694–1697. [CrossRef] [PubMed]
9. Barthod, J.; Rumpel, C.; Dignac, M.F. Composting with additives to improve organic amendments. A review. *Agron. Sustain. Dev.* **2018**, *38*, 17. [CrossRef]
10. Van Zwieten, L. The long-term role of organic amendments in addressing soil constraints to production. *Nutr. Cycl. Agroecosyst.* **2018**, *111*, 99–102. [CrossRef]
11. Tilman, D.; Reich, P.B.; Knops, J.M. Biodiversity and ecosystem stability in a decadelong grassland experiment. *Nature* **2006**, *441*, 629–632. [CrossRef]
12. Ryan, L.F.; Daniel, B.R.; Dharni, V.; Mark, A.W.; Larry, T. Rhizogenic Fe-C redox cycling: A hypothetical biogeochemical mechanism that drives crustal weathering in upland soils. *Biogeochemistry* **2008**, *87*, 127–141.
13. Poulsen, P.H.; Al-Soud, W.A.; Bergmark, L.; Magid, J.; Hansen, L.H.; Sørensen, S.J. Effects of fertilization with urban and agricultural organic wastes in a field trial–Prokaryotic diversity investigated by pyrosequencing. *Soil Biol. Biochem.* **2013**, *57*, 784–793. [CrossRef]
14. Yuan, H.; Ge, T.; Zhou, P.; Liu, S.; Roberts, P.; Zhu, H.; Zou, Z.; Tong, C.; Wu, J. Soil microbial biomass and bacterial and fungal community structures responses to long-term fertilization in paddy soils. *J. Soils Sediments* **2013**, *13*, 877–886. [CrossRef]
15. Biswas, B.; Pandey, N.; Bisht, Y.; Singh, R.; Kumar, J.; Bhaskar, T. Pyrolysis of agricultural biomass residues: Comparative study of corn cob, wheat straw, rice straw and rice husk. *Bioresour. Technol.* **2017**, *237*, 57–63. [CrossRef] [PubMed]
16. Cui, Y.F.; Meng, J.; Wang, Q.X.; Zhang, W.M.; Cheng, X.Y.; Chen, W.F. Effects of straw and biochar addition on soil nitrogen, carbon, and super rice yield in cold waterlogged paddy soils of North China. *J. Integr. Agric.* **2017**, *16*, 1064–1074. [CrossRef]
17. Xia, L.; Lam, S.K.; Yan, X.; Chen, D. How does recycling of livestock manure in agroecosystems affect crop productivity, reactive nitrogen losses, and soil carbon balance? *Environ. Sci. Technol.* **2017**, *51*, 7450–7457. [CrossRef] [PubMed]
18. Samaddar, S.; Truu, J.; Chatterjee, P.; Truu, M.; Kim, K.; Kim, S.; Seshadri, S.; Sa, T. Long-term silicate fertilization increases the abundance of Actinobacterial population in paddy soils. *Biol. Fertil. Soils* **2019**, *55*, 109–120. [CrossRef]
19. Liu, N.; Gumpertz, M.L.; Hu, S.; Ristaino, J.B. Long-term effects of organic and synthetic soil fertility amendments on soil microbial communities and the development of southern blight. *Soil Biol. Biochem.* **2007**, *39*, 2302–2316. [CrossRef]
20. Huse, S.M.; Dethlefsen, L.; Huber, J.A.; Welch, D.M.; Relman, D.A.; Sogin, M.L. Exploring microbial diversity and taxonomy using SSU rRNA hypervariable tag sequencing. *PLoS Genet.* **2008**, *4*, 1000255. [CrossRef]
21. Ruijter, J.M.; Ramakers, C.; Hoogaars, W.M.H.; Karlen, Y.; Bakker, O.; Van den Hoff, M.J.B.; Moorman, A.F.M. Amplification efficiency: Linking baseline and bias in the analysis of quantitative PCR data. *Nucleic Acids Res.* **2009**, *37*, e45. [CrossRef]
22. Schloss, P.D.; Westcott, S.L.; Ryabin, T.; Hall, J.R.; Hartmann, M.; Hollister, E.B.; Lesniewski, R.A.; Oakley, B.B.; Parks, D.H.; Robinson, C.J.; et al. Introducing mothur: Opensource, platform-independent, community supported software for describing and comparing microbial communities. *Appl. Environ. Microbiol.* **2009**, *75*, 7537–7541. [CrossRef] [PubMed]

23. Samaddar, S.; Chatterje, P.; Truu, J.; Anandham, R.; Kim, S.; Sa, T. Long-term phosphorus limitation changes the bacterial community structure and functioning in paddy soils. *Appl. Soil Ecol.* **2019**, *134*, 111–115. [CrossRef]

24. SAS Institute Inc. *Base Sas 9.4 Procedures Guide: Statistical Procedures*, 2nd ed.; SAS Institute Inc.: Cary, NC, USA, 2013.

25. Segata, N.; Izard, J.; Waldron, L.; Gevers, D.; Miropolsky, L.; Garrett, W.S.; Huttenhower, C. Metagenomic biomarker discovery and explanation. *Genome Biol.* **2011**, *12*, 1–18. [CrossRef]

26. Langille, M.G.I.; Zaneveld, J.; Caporaso, J.G.; McDonald, D.; Knights, D.; Reyes, J.A.; Clemente, J.C.; Burkepile, D.E.; Vega Thurber, R.L.; Knight, R.; et al. Predictive functional profiling of microbial communities using 16S rRNA marker gene sequences. *Nat. Biotechnol.* **2013**, *31*, 814–821. [CrossRef] [PubMed]

27. Li, W.X.; Wang, C.; Zheng, M.M.; Cai, Z.J.; Wang, B.R.; Shen, R.F. Fertilization strategies affect soil properties and abundance of N-cycling functional genes in an acidic agricultural soil. *Appl. Soil Ecol.* **2020**, *156*, 103704. [CrossRef]

28. Calleja-Cervantes, M.E.; Fernández-González, A.J.; Irigoyen, I.; Fernández-López, M.; Aparicio-Tejo, P.M.; Menéndez, S. Thirteen years of continued application of composted organic wastes in a vineyard modify soil quality characteristics. *Soil Biol. Biochem.* **2015**, *90*, 241–254. [CrossRef]

29. Tian, W.; Wang, L.; Li, Y.; Zhuang, K.; Li, G.; Zhang, J.; Xiao, X.; Xi, Y. Responses of microbial activity, abundance, and community in wheat soil after three years of heavy fertilization with manure-based compost and inorganic nitrogen. *Agric. Ecosyst. Environ.* **2015**, *213*, 219–227. [CrossRef]

30. Hartmann, M.; Widmer, F. Community structure analyses are more sensitive to differences in soil bacterial communities than anonymous diversity indices. *Appl. Environ. Microbiol.* **2006**, *72*, 7804–7812. [CrossRef] [PubMed]

31. Shade, A. Diversity is the question, not the answer. *ISME J.* **2017**, *11*, 1–6. [CrossRef]

32. Janssen, P.H. Identifying the dominant soil bacterial taxa in libraries of 16S rRNA and 16S rRNA genes. *Appl. Environ. Microbiol.* **2006**, *72*, 1719–1728. [CrossRef]

33. Spain, A.M.; Krumholz, L.R.; Elshahed, M.S. Abundance, composition, diversity and novelty of soil Proteobacteria. *ISME J.* **2009**, *3*, 992–1000. [CrossRef]

34. Zhalnina, K.; Dias, R.; de Quadros, P.D.; Davis-Richardson, A.; Camargo, F.A.; Clark, I.M.; McGrath, S.P.; Hirsch, P.R.; Triplett, E.W. Soil pH determines microbial diversity and composition in the park grass experiment. *Microb. Ecol.* **2015**, *69*, 395–406. [CrossRef]

35. Zhao, Y.; Wang, P.; Li, J.; Chen, Y.; Ying, X.; Liu, S. The effects of two organic manures on soil properties and crop yields on a temperate calcareous soil under a wheat-maize cropping system. *Eur. J. Agron.* **2009**, *31*, 36–42. [CrossRef]

36. Tanahashi, T.; Murase, J.; Matsuya, K.; Hayashi, M.; Kimura, M.; Asakawa, S. Bacterial communities responsible for the decomposition of rice straw compost in a Japanese rice paddy field estimated by DGGE analysis of amplified 16S rDNA and 16S rRNA fragments. *Soil Sci. Plant Nutr.* **2005**, *51*, 351–360. [CrossRef]

37. Sharmin, F.; Wakelin, S.; Huygens, F.; Hargreaves, M. Firmicutes dominate the bacterial taxa within sugar-cane processing plants. *Sci. Rep.* **2013**, *3*, 3107. [CrossRef] [PubMed]

38. Dai, Z.; Su, W.; Chen, H.; Barberán, A.; Zhao, H.; Yu, M.; Yu, L.; Brookes, P.C.; Schadt, C.W.; Chang, S.X.; et al. Long-term nitrogen fertilization decreases bacterial diversity and favors the growth of Actinobacteria and Proteobacteria in agro-ecosystems across the globe. *Glob. Change Biol.* **2018**, *24*, 3452–3461. [CrossRef]

39. Tsoy, O.V.; Ravcheev, D.A.; Čuklina, J.; Gelfand, M.S. Nitrogen fixation and molecular oxygen: Comparative genomic reconstruction of transcription regulation in Alphaproteobacteria. *Front. Microbiol.* **2016**, *7*, 1343. [CrossRef] [PubMed]

40. Douglas, G.M.; Beiko, R.G.; Langille, M.G. Predicting the functional potential of the microbiome from marker genes using PICURSt. In *Microbiome Analysis*; Beiko, R., Hsiao, W., Parkinson, J., Eds.; Humana Press: New York, NY, USA, 2018; Volume 1849, pp. 169–177.

41. Pii, Y.; Borruse, L.; Brusetti, L.; Crecchio, C.; Cesco, S.; Mimmo, T. The interaction between iron nutrition, plant species and soil type shapes the rhizosphere microbiome. *Plant Physiol. Biochem.* **2016**, *99*, 39–48. [CrossRef]

42. Zarraonaindia, I.; Owens, S.M.; Weisenhorn, P.; West, K.; Hampton-Marcell, J.; Lax, S.; Bokulich, N.A.; Mills, D.A.; Martin, G.; Taghavi, S.; et al. The soil microbiome influences grapevine-associated microbiota. *mBio* **2015**, *6*, e02527-14. [CrossRef] [PubMed]

43. Dube, J.P.; Valverde, A.; Steyn, J.M.; Cowan, D.A.; Van der Walls, J.E. Differences in bacterial diversity, composition and function due to long-term agriculture in soils in the eastern free State of South Africa. *Diversity* **2019**, *11*, 61. [CrossRef]

44. Zhong, Y.; Yang, Y.; Liu, P.; Xu, R.; Rensing, C.; Fu, X.; Liao, H. Genotype and rhizobium inoculation modulate the assembly of soybean rhizobacterial communities. *Plant Cell Environ.* **2019**, *42*, 2028–2044. [CrossRef]

45. Glick, B.R. Bacteria with ACC deaminase can promote plant growth and help to feed the world. *Microbiol. Res.* **2014**, *169*, 30–39. [CrossRef] [PubMed]

46. Pollmann, S.; Neu, D.; Weiler, E.W. Molecular cloning and characterization of an amidase from Arabidopsis thaliana capable of converting indole-3-acetamide into the plant growth hormone, indole-3-acetic acid. *Phytochemistry* **2003**, *62*, 293–300. [CrossRef]

47. Koyama, H.; Kawamura, A.; Kihara, T.; Hara, T.; Takita, E.; Shibata, D. Overexpression of mitochondrial citrate synthase in Arabidopsis thaliana improved growth on a phosphorus-limited soil. *Plant Cell Physiol.* **2000**, *41*, 1030–1037. [CrossRef] [PubMed]

48. Barone, P.; Rosellini, D.; LaFayette, P.; Bouton, J.; Veronesi, F.; Parrott, W. Bacterial citrate synthase expression and soil aluminum tolerance in transgenic alfalfa. *Plant Cell Rep.* **2008**, *27*, 893–901. [CrossRef] [PubMed]

49. López-Mondéjar, R.; Zühlke, D.; Becher, D.; Riedel, K.; Baldrian, P. Cellulose and hemicellulose decomposition by forest soil bacteria proceeds by the action of structurally variable enzymatic systems. *Sci. Rep.* **2016**, *6*, 25279. [CrossRef]

50. Han, W.; He, M. The application of exogenous cellulase to improve soil fertility and plant growth due to acceleration of straw decomposition. *Bioresour. Technol.* **2010**, *101*, 3724–3731. [CrossRef] [PubMed]
51. Tiquia, S.M.; Tam, N.F.Y.; Hodgkiss, I.J. Microbial activities during composting of spent pig-manure sawdust litter at different moisture contents. *Bioresour. Technol.* **1996**, *55*, 201–206. [CrossRef]
52. Sunkar, R.; Bartels, D.; Kirch, H.H. Overexpression of a stress-inducible aldehyde dehydrogenase gene from Arabidopsis thaliana in transgenic plants improves stress tolerance. *Plant J.* **2003**, *35*, 452–464. [CrossRef]
53. Schnurr, J.; Shockey, J. The acyl-CoA synthetase encoded by LACS2 is essential for normal cuticle development in Arabidopsis. *Plant Cell* **2004**, *16*, 629–642. [CrossRef] [PubMed]
54. Wang, X.L.; Li, X.B. The GhACS1 gene encodes an acyl-CoA synthetase which is essential for normal microsporogenesis in early anther development of cotton. *Plant J.* **2009**, *57*, 473–486. [CrossRef] [PubMed]
55. De Azevedo Souza, C.; Kim, S.S.; Koch, S.; Kienow, L.; Schneider, K.; McKim, S.M.; Haughn, G.W.; Kombrink, E.; Douglas, C.J. A novel fatty acyl-CoA synthetase is required for pollen development and sporopollenin biosynthesis in Arabidopsis. *Plant Cell* **2009**, *21*, 507–525. [CrossRef] [PubMed]
56. O'Hara, P.; Slabas, A.R.; Fawcett, T. Antisense expression of 3-oxoacyl-ACP reductase affects whole plant productivity and causes collateral changes in activity of fatty acid synthase components. *Plant Cell Physiol.* **2007**, *48*, 736–744. [CrossRef] [PubMed]

Communication

Correlation between Soil Bacterial Community Structure and Soil Properties in Cultivation Sites of 13-Year-Old Wild-Simulated Ginseng (*Panax ginseng* C.A. Meyer)

Kiyoon Kim, Hyun Jun Kim, Dae Hui Jeong, Jeong Hoon Huh, Kwon Seok Jeon and Yurry Um *

Forest Medicinal Resources Research Center, National Institute of Forest Science, Yongju 36040, Korea; kky0607@korea.kr (K.K.); mind4938@korea.kr (H.J.K.); najdhda@korea.kr (D.H.J.); mdgs3275@korea.kr (J.H.H.); jks2029@korea.kr (K.S.J.)
* Correspondence: urspower@korea.kr

Abstract: Soil properties are one of the major factors determining the growth of vegetation. These properties drive the selection of the dominant bacterial community profiles, which eventually determines the soil quality and fertility. The abundance of preferential bacterial community assists in better productivity of a particular type of vegetation. The increasing focus on the health and well-being of the human population has resulted in a shift in paradigm to concentrate on the cultivation of medicinal plants such as Wild-simulated ginseng (WSG). These plant species take a long time for their growth and are generally cultivated in the mountainous forest trenches of Far East countries like South Korea. This study was conducted to decipher the bacterial community profiles and their correlation with soil chemical properties, which would give a broader idea about the optimum growing conditions of such an important medicinal plant. The important edaphic factor determined in this study was the soil pH, which was recorded to be acidic in all the studied cultivation sites. In agreement with the edaphic factor, the relative abundance of *Acidobacteria* was found to be highest as this phylum prefers to grow in acidic soils. Moreover, the total organic matter, total nitrogen and cation exchange capacity were found to be significantly correlated with the bacterial community. Hence, these results will help to identify the suitable cultivation sites for WSG and increase the productivity of these medicinal plants.

Keywords: wild-simulated ginseng; *Panax ginseng* C.A. Meyer; soil bacterial community; soil property; correlation analysis

Citation: Kim, K.; Kim, H.J.; Jeong, D.H.; Huh, J.H.; Jeon, K.S.; Um, Y. Correlation between Soil Bacterial Community Structure and Soil Properties in Cultivation Sites of 13-Year-Old Wild-Simulated Ginseng (*Panax ginseng* C.A. Meyer). *Appl. Sci.* **2021**, *11*, 937. https://doi.org/10.3390/app11030937

Received: 21 December 2020
Accepted: 18 January 2021
Published: 20 January 2021

Publisher's Note: MDPI stays neutral with regard to jurisdictional claims in published maps and institutional affiliations.

1. Introduction

Wild-simulated ginseng (WSG) belongs to the Araliaceae family, and it is also known as *Panax ginseng* C.A. Meyer [1]. It is mostly grown through artificially sowing the seeds or transplanting of seedlings in a mountainous area by the Korean Forest Service (KFS) [2]. In Korea, the WSG is defined as a kind of ginseng produced without the use of any artificial facilities, and the West Virginia legislature in the United States defined it as the ginseng grown in theforest without the use of any weed, disease or pest control agents. [3].

Soil microbes present in the rhizosphere have symbiotic relations with plants, and they can contribute to plant growth through decomposition of organic matter, nutrient (carbon, nitrogen and inorganic elements) cycling, removal of pollutants and supplying of nutrients to plants, and they play an important role in determining soil quality and productivity [4,5]. The recent develop in culture-independent methods has made it convenient to study microbial diversity and predict key functional traits of soil microbiota [6]. Myriad environmental factors can affect the soil properties and in turn tweak diversity and composition of soil microbiota [7,8]. Therefore, studying the correlation of environmental factors and soil microbial community is very important [9–14].

The recent focus on health and immunity has enhanced the interest in organic agriculture and exploitation of soil microbes for improvement in quality, and productivity of medicinal crops is pacing up [1,15]. Ginseng (*Panax ginseng*) is a representative medicinal crop used in Far East countries, and there has been increasing interest in studying the soil microbial community present in their cultivation fields [16,17]. In addition, there has been a growing interest in studying the soil microbial communities based on the changes in forest environments [18–20]. However, the correlation between soil microbial communities and edaphic factors exerted on medicinal crops growing in forest is insufficiently studied.

The correlation of the soil microbial community with the edaphic factors for cultivation of WSG is important, as it is cultivated in the mountainous trench for a long period of time (~7–15 years) without the use of any pesticide or chemical fertilizers [21]. Thus, before cultivating WSG, it is necessary to investigate the suitability of cultivation by analyzing the edaphic factors of the site, such as soil properties and soil microbial communities [3]. Hence, the aim of this study was to investigate the correlation between soil properties and soil bacterial communities in different cultivation sites of WSG grown for 13 years in the forest ecosystem.

2. Materials and Methods

2.1. Study Area and Soil Sample Collection

A total of 9 different cultivation sites of WSG were chosen randomly in South Korea, and the details of the study area and the sampling sites are shown in Figure 1. Both rhizospheric and non-rhizospheric soil samples were collected in three replicates from each cultivation site from July to August in 2019. The rhizosphere soil was stored at −20 °C for analysis of soil bacterial community, and non-rhizosphere soil was sieved and air-dried for analysis of soil chemical properties. The characteristic features of the cultivation site of wild-simulated ginseng were recorded by studying the usual forest physiognomy such as tree species, tree height (TH) and diameter of breast height (DBH), and the topography such as slope direction, slope gradient and height above sea level (HASL) within the stipulated 10 m × 10 m plots of each cultivation site.

Figure 1. Study area and soil sampling sites of 13-year-old Wild-simulated ginseng (WSG). A-I are a wild-simulated ginseng(WSG) cultivation sites in South Korea.

2.2. Soil Analysis

Surface soil was removed, and soil was collected at a depth within 20 cm. The soil samples were passed through a 2 mm sieve and air-dried at room temperature. Soil chemical properties analysis was performed following standard analysis manual of the Rural Development Administration (RDA), South Korea [22].

2.3. Soil DNA Extraction and PCR Amplification

Total DNA of each rhizosphere soil samples was extracted using DNeasy PowerSoil kit (QIAGEN, Hilden, Germany) following manufacturer instructions. After extraction, quantification and quality of DNA were measured by PicoGreen and Nanodrop (Thermo Scientific, Rockford, IL, USA). Each sequenced sample was prepared according to the Illumina 16S Metagenomics Sequencing Library protocols (Macrogen, Seoul, Korea). In amplicon PCR, V3-V4 region of the 16S rRNA gene of bacteria was targeted using the 16S V3-V4 primers [23]. The 16S V3-V4 primer sequences are as follows: 16S amplicon PCR forward primer, 5′- TCGTCGGCAGCGTCAGATGTGTATAAGAGACAGCCTACGGGNGGCWGC AG-3′, 16S amplicon PCR reverse primer, 5′-GTCTCGTGGGCTCGGAGATGTGTATAAGA GACAGGACTACHVGGGTATCTAATCC -3′. Input gDNA was amplified with 16S V3-V4 primers, and a subsequent limited-cycle amplification step was performed to add multiplexing indices and Illumina sequencing adapters. The conditions for amplicon PCR were as follows. First PCR: initial denaturation at 95 °C for 3 min, followed by 25 cycles of denaturation at 95 °C for 30 s, annealing at 55 °C for 30 s and extension at 72 °C for 30 s, and a final extension at 72 °C for 5 min. The condition for index PCR was as follows. Second PCR: initial denaturation at 95 °C for 3 min, followed by 8 cycles of denaturation at 95 °C for 30 s, annealing at 55 °C for 30 s, and extension at 72 °C for 30 s, and a final extension at 72 °C for 5 min. The final products were normalized and pooled using the PicoGreen, and the size of libraries were verified using the TapeStation DNA screentape D1000 (Agilent, Santa Clara, CA, USA).

2.4. Pyrosequencing and Data Processing

Bacterial DNA sequencing was performed using the Illumina MiSeq™ sequencing system (Illumina Inc., San Diego, CA, USA) according to the manufacturer's instructions. Raw sequences of bacterial DNA were processed using Mothur pipeline (version 1.43.0, The University of Michigan, Ann Arbor, MI, USA) [24]. The forward and reverse reads obtained from Illumina platform were assembled, and sequences with the quality score of <20 and the ambiguous nucleotides were discarded before performing downstream analysis. The resulting sequences spanning the V3-V4 region were checked for the presence of chimera using the function chimera.uchime. Taxonomic classification was performed using the "Greengenes reference database" for bacteria. Greengenes was used as it was reported to provide the best combination of speed and quality [25]. The sequences were clustered into operational taxonomic units (OTUs) at 97% similarity level using distance-based greedy clustering method (DGC) in Mothur. OTUs with less than 10 sequences were discarded to reduce false diversity.

2.5. Data Analysis

Data are expressed as means ± standard error (S.E.). Statistical analysis was performed using the program Statistical Analysis System (SAS, version 9.4, SAS Institute, Cary, NC, USA) software for one-way ANOVA and Duncan's test, with statistical significance set at $p < 0.05$ [26]. The data analysis and processing of the 16S rRNA amplicon data was performed following the guidelines [27]. The richness estimators (ACE, Chao and Jackknife) and alpha diversity indices (Shannon and Inverted Simpson) were calculated using Mothur. The principal coordinate analysis (PCoA) was performed using Mothur to visualize the relationship with soil factors based on bacterial community composition. Differences in bacterial community composition were tested using Bray-Curtis dissimilarity values with permutational analysis of variance (PERMANOVA), which is a nonparametric technique used to differentiate groups based on dissimilarity matrix [28]. DistLM program with 10,000 permutations was used to identify the soil factors explaining the variations in community structure. Correlation coefficient analysis between soil factors and diversity indices were analyzed using Pearson's correlation (IBM SPSS Statistics, version 25, IBM Corp., Armonk, NY, USA).

3. Results and Discussion

3.1. Location Environment (Topography, Forest Physiognomy, Soil Properties) of the Study Area

The topography and forest physiognomy of WSG cultivation sites are summarized in Table 1. In general, all the cultivation sites were sloped terrain with slope gradient ranging from 5 to 35°; the slope direction varied from east, north, southeast, southwest and northeast; and the sites were 330–920 m above sea level. On the other hand, D and E cultivation sites were identified as broad-leaved forest, and all other cultivation sites were identified as mixed forest of conifer and broad-leaved. Among the cultivation sites, the average TH was maximum in the F cultivation site (26.8 m), and the average DBH was maximum in the A cultivation site (36.1 cm). Furthermore, the chemical properties of soil samples are summarized in Table 2. Soil samples were classified as sandy loam and sandy clay loam based on their soil texture. The soil pH of all cultivation sites has been recorded as acidic soil, and the I cultivation site showed the significantly lowest value compared to other cultivation sites. Organic matter (OM), total nitrogen (TN) contents and cation exchange capacity (CEC) were significantly higher in the A cultivation site, whereas the available phosphate (avail. P_2O_5) content was significantly higher in the B cultivation site compared to other cultivation sites. Furthermore, potassium (K) content was recorded at a range of 0.08 to 0.31 cmol$^+$ kg^{-1}, calcium (Ca) in the range of 0.10 to 6.99 cmol$^+$ kg^{-1}, magnesium (Mg) in the range of 0.05 to 1.07 cmol$^+$ kg^{-1} and sodium (Na) in the range of 0.03 to 0.09 cmol$^+$ kg^{-1}, which belongs to the group of exchangeable ions. Forest vegetation is formed by interaction with the environment, and among the forest environments, soil characteristics are majorly affected by the vegetation, and it varies significantly according to the difference in the presence of the particular species of trees [29]. Therefore, the growth and production of WSG cultivated in forest regions have a significant correlation with forest soil and tree species [30]. The organic matter content is higher in broad-leaved forests than in coniferous forests in forest soil because the accumulation of fallen leaves from the trees determines the organic matter content [31,32]. Among the WSG cultivation sites, soil organic matter and total nitrogen content are significantly high in mixed forests with high diversity of deciduous broad-leaved trees [33]. This is because broad-leaved forests contain more organic carbon sources such as fallen leaves than coniferous forests, where organic matter is slowly decomposed [34]. In this study, OM, TN and CEC were high in the WSG cultivation sites with a high percentage broad-leaved tree.

Table 1. Location environments of 13-year-old WSG cultivation sites.

Cultivation Sites	Topography			Forest Physiognomy			
	Slope		HASL [a]	Species of Tree	Average		Percentage
					TH [b]	DBH [c]	
	°	Detection	m		m	cm	%
A	32	Southeast	920	Broad-leaved	21.5	31.0	80.0
				Conifer	36.0	56.7	20.0
				Total	24.4	36.1	100
B	20	Southwest	615	Broad-leaved	22.8	14.8	35.7
				Conifer	19.0	13.0	64.3
				Total	20.4	13.7	100
C	30	Northeast	387	Broad-leaved	14.7	12.8	81.8
				Conifer	31.5	29.6	18.2
				Total	17.8	15.9	100

Table 1. *Cont.*

Cultivation Sites	Topography			Forest Physiognomy			
	Slope		HASL [a]	Species of Tree	Average		Percentage
					TH [b]	DBH [c]	
	°	Detection	m		m	cm	%
D	20	North	530	Broad-leaved	18.8	15.0	100
				Conifer	-	-	-
				Total	18.8	15.0	100
E	27	Southwest	330	Broad-leaved	16.2	15.1	100
				Conifer	-	-	-
				Total	16.2	15.1	100
F	35	Southeast	717	Broad-leaved	27.3	22.8	90.9
				Conifer	22.0	23.8	9.1
				Total	26.8	22.9	100
G	25	East	712	Broad-leaved	17.1	21.6	40.0
				Conifer	27.0	35.6	60.0
				Total	21.7	28.0	100
H	15	North	743	Broad-leaved	14.5	18.2	40.0
				Conifer	27.0	34.2	60.0
				Total	22.0	27.8	100
I	5	Southeast	406	Broad-leaved	21.0	24.8	50.0
				Conifer	20.0	36.6	50.0
				Total	20.5	30.7	100

[a] HSAL: height above sea level; [b] TH: tree height; [c] DBH: diameter of breast height.

3.2. Bacterial Community Profiles

The bacterial community profiles varied among the soil samples of 13-year-old WSG cultivation sites. The relative abundance of bacterial community at phylum levels is shown in Figure 2. The soil bacterial communities were grouped based on the cultivation sites. *Acidobacteria* (33.6%) was the most dominant phylum in all soil samples, followed by *Proteobacteria* (23.9%), *Verrucomicrobia* (11.2%), *Chloroflexi* (5.9%), *Actinobacteria* (4.4%) and *Planctomycetes* (3.9%). The relative abundance of *Acidobacteria* and *Chloroflexi* was significantly higher in cultivation site I compared to the other cultivation sites, whereas that of *Proteobacteria* was significantly higher in cultivation site F. On the other hand, *Verrucomicrobia*, *Actinobacteria* and *Plantomycetes* were significantly more abundant in cultivation site C. This observation corroborates to previous studies where *Acidobacteria*, *Proteobacteria*, *Verrucomicrobia* and *Actinobacteria* were the major bacterial communities at the phylum level present in soils used for cultivation of *Panax ginseng* [3,35,36]. *Acidobacteria* are acidophilic bacteria mainly present in acidic soils [37]; hence, soil pH is one of the major factors determining *Acidobacteria's* community composition [38–40]. Bacteria belonging to *Acidobacteria* have evolved mechanisms that prefer acidic pH by stabilization of intracellular enzymes [41]. In this study, the relative abundance of *Acidobacteria* was shown to be significantly higher in the cultivation site I, which had the lowest soil pH compared to other studied groups. *Acidobacteria* are shown to be negatively correlated with soil pH in WSG cultivation sites [3], and it was also reported that the *Acidobacteria* population is higher in the cultivated soil of *Panax ginseng* [42,43].

Table 2. Soil chemical properties of the samples from 9 different cultivation sites of WSG.

Cultivation Sites	Soil Texture	pH	EC [a]	OM [b]	TN [c]	Avail. P_2O_5 [d]	Exchangeable Cation				CEC [e]
							K	Ca	Mg	Na	
		(1:5)	dS m^{-1}	%	%	mg kg^{-1}	cmol$^+$ kg^{-1}	cmol$^+$ kg^{-1}	cmol$^+$ kg^{-1}	cmol$^+$ kg^{-1}	cmol$^+$ kg^{-1}
A	Sandy clay loam	4.91 ± 0.07 [ab]	0.03 ± 0.01 [a]	17.2 ± 1.18 [a]	0.69 ± 0.06 [a]	18.1 ± 4.4 [d]	0.18 ± 0.08 [ab]	1.36 ± 0.74 [cd]	0.35 ± 0.16 [b]	0.09 ± 0.05 [a]	36.6 ± 5.2 [a]
B	Sandy loam	5.03 ± 0.07 [a]	0.02 ± 0.00 [a]	3.8 ± 0.41 [bc]	0.15 ± 0.01 [c]	149.6 ± 9.0 [a]	0.11 ± 0.01 [b]	1.76 ± 0.15 [d]	0.28 ± 0.04 [b]	0.04 ± 0.00 [a]	14.3 ± 2.2 [b]
C	Sandy clay loam	5.61 ± 0.14 [a]	0.02 ± 0.00 [a]	11.6 ± 0.56 [d]	0.42 ± 0.01 [d]	8.6 ± 0.3 [d]	0.15 ± 0.04 [ab]	4.70 ± 0.39 [b]	1.06 ± 0.02 [a]	0.08 ± 0.02 [a]	30.9 ± 2.1 [c]
D	Sandy loam	5.61 ± 0.35 [ab]	0.02 ± 0.00 [a]	9.5 ± 1.00 [bc]	0.32 ± 0.05 [bc]	26.2 ± 8.9 [d]	0.08 ± 0.01 [b]	0.54 ± 0.27 [cd]	0.11 ± 0.06 [b]	0.06 ± 0.01 [a]	27.1 ± 2.7 [ab]
E	Sandy clay loam	5.29 ± 0.05 [ab]	0.05 ± 0.01 [a]	12.7 ± 0.21 [d]	0.48 ± 0.01 [d]	74.1 ± 13.4 [bc]	0.30 ± 0.04 [a]	6.99 ± 0.53 [d]	1.07 ± 0.37 [b]	0.07 ± 0.02 [a]	33.5 ± 0.4 [c]
F	Sandy loam	5.11 ± 0.04 [ab]	0.03 ± 0.00 [a]	8.5 ± 0.24 [c]	0.34 ± 0.04 [c]	105.8 ± 23.2 [b]	0.19 ± 0.03 [ab]	2.72 ± 0.77 [c]	0.38 ± 0.11 [b]	0.06 ± 0.01 [a]	27.9 ± 0.9 [b]
G	Sandy clay loam	5.16 ± 0.07 [ab]	0.02 ± 0.00 [a]	8.7 ± 0.61 [c]	0.34 ± 0.01 [c]	60.5 ± 10.6 [c]	0.17 ± 0.02 [ab]	2.03 ± 0.19 [cd]	0.39 ± 0.11 [b]	0.05 ± 0.01 [a]	27.6 ± 0.4 [b]
H	Sandy loam	5.06 ± 0.20 [ab]	0.05 ± 0.00 [a]	4.3 ± 0.88 [b]	0.17 ± 0.04 [b]	14.8 ± 3.8 [d]	0.31 ± 0.01 [a]	0.55 ± 0.16 [a]	0.16 ± 0.03 [a]	0.03 ± 0.01 [a]	14.7 ± 2.2 [ab]
I	Sandy loam	4.73 ± 0.04 [b]	0.02 ± 0.00 [a]	8.7 ± 1.06 [c]	0.34 ± 0.04 [c]	23.2 ± 9.6 [d]	0.11 ± 0.02 [b]	0.10 ± 0.02 [d]	0.05 ± 0.01 [b]	0.06 ± 0.02 [a]	26.0 ± 2.2 [b]

Each column shows the means of three replications ± standard error (SE). Values in each column with different letters show statistically significant differences ($p < 0.05$) among the treatments according to Duncan's test. [a] EC: electrical conductivity; [b] OM: organic matter; [c] TN: total nitrogen; [d] Avail. P2O5: available phosphorus, [e] CEC: cation exchange capacity.

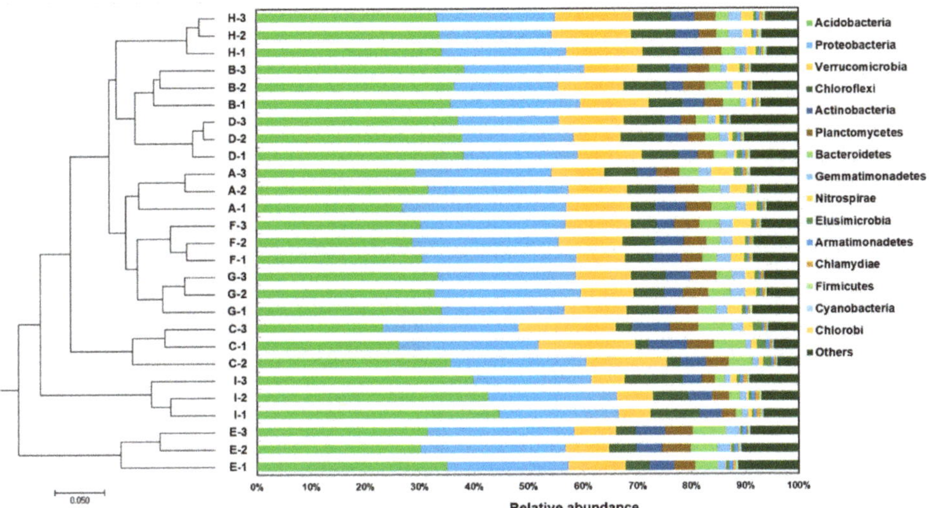

Figure 2. The relative abundance of taxonomic composition at the phylum level for bacteria across soil samples, with clustering tree on the left showing the similarities among the soil samples.

3.3. Correlation between Soil Bacterial Community and Soil Properties

Principal coordinate analysis (PCoA) and DistLM was done to analyze the correlation between soil bacterial community and the edaphic factors. The two axes of PCoA (Figure 3) explained 45.4% of the total variation in the bacterial community, and soil factors located at each coordinate are related to soil bacterial communities divided by ordinate or abscissa. Soil factors placed in the abscissa are more correlated with soil bacterial communities compared to those located in the ordinate coordinates because the PC1 variation (27.7%) is higher than the variation of PC2 (17.7%). In other words, soil OM, TN, CEC, Mg and Ca affects soil bacterial clustering more than the other soil factors. The DistLM analysis indicated significant correlation between soil factors and soil bacterial community. Cation exchange capacity (CEC), OM, TN, P_2O_5, Mg and K were significantly affecting the bacterial community (Table 3). Regarding the sequential tests, the CEC, OM, TN, pH and P_2O_5 had a more significant effect on the bacterial community compared to other soil factors. The results of Pearson's correlation analysis between soil factors and diversity indices of bacterial community are represented in Table 4. Among the diversity indices, ace, chao and Shannon diversity index were shown to have a significant positive correlation with OM, TN and CEC. The correlation between soil microbial communities and soil properties had been carried out in numerous studies [44,45]. A study that concentrated on studying the correlation between soil properties and bacterial community in WSG cultivation sites showed that the soil bacterial community is significantly correlated with soil pH, OM, TN and CEC [36]. Soil microorganisms inhabiting the soil have an important relationship with soil quality and productivity such as OM decomposition and nutrient cycling. In addition, the decomposition of OM and nitrogen mineralization in the soil proceeds through a complex interaction of abiotic factors such as soil properties and biotic factors such as microbial population and nutrient demand [46,47]. Soil microorganisms are an important factor affecting soil fertility [48]. The cation exchangeable capacity (CEC) is an indicator of soil fertility and is involved in improving soil buffer capacity, nutrient holding capacity and supplying nutrients [49]. In general, OM, TN and CEC have a high correlation in the natural vegetation [50]. In the results of this study, the soil bacterial community had a significant correlation with OM, TN and CEC, and this is considered to have a significant correlation with the growth characteristics of WSG.

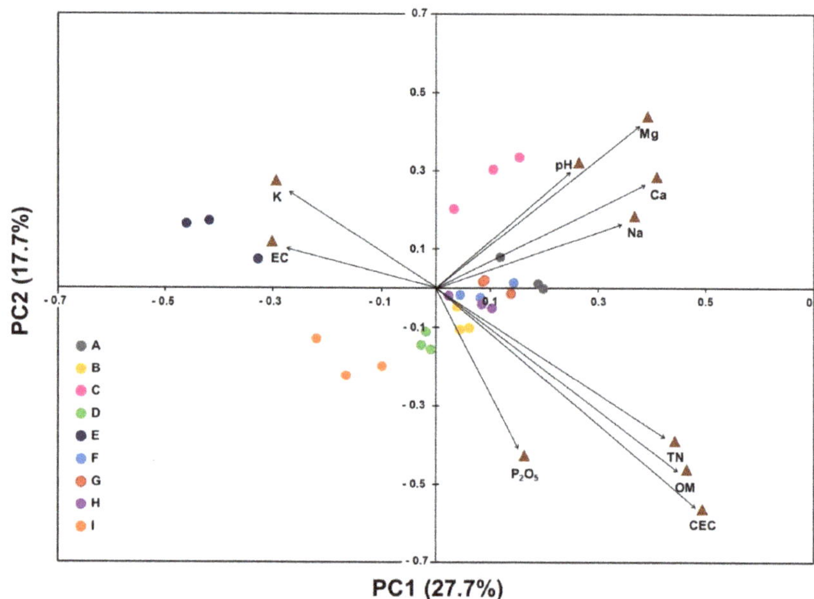

Figure 3. Principal coordinate analysis based on Bray–Curtis dissimilarity matrix for the bacterial community generated using themothur platform.

Table 3. Marginal and sequential tests of DISTLM on the relation of soil factors variables to the bacterial community of soil samples.

Soil Factors	Marginal Test		Sequential Test		
	p-Value	Proportion	*p*-Value	Proportion	Cumulative
CEC	0.0003	0.1410	0.0003	0.1410	0.1410
OM	0.0011	0.1227	0.0002	0.1089	0.2499
TN	0.0040	0.1095	0.0017	0.1026	0.3525
P_2O_5	0.0425	0.0721	0.0269	0.0522	0.4047
pH	0.0627	0.0669	0.0297	0.0487	0.4535
Mg	0.0078	0.1025	0.3088	0.0290	0.4824
K	0.0522	0.0684	0.1213	0.0367	0.5191
Na	0.2844	0.0618	0.5460	0.0230	0.5421
EC	0.1731	0.0509	0.6749	0.0200	0.5620
Ca	0.2113	0.0945	0.6489	0.0208	0.5828

Table 4. Pearson's correlation analysis between soil factors and diversity indices in cultivation sites of WSG.

Soil Factors	Correlation Coefficient (r)				
	Ace	Chao	Jackknife	Shannon	Invsimpson
pH	0.225 (0.260)	0.229 (0.251)	−0.188 (0.348)	0.228 (0.253)	0.122 (0.544)
EC	−0.074 (0.714)	−0.236 (0.236)	−0.132 (0.512)	−0.234 (0.240)	0.016 (0.938)
OM	0.459 * (0.016)	0.380 * (0.050)	−0.115 (0.569)	0.407 * (0.035)	−0.246 (0.216)
TN	0.425 * (0.027)	0.371 (0.057)	−0.118 (0.557)	0.391 * (0.043)	−0.166 (0.408)
P_2O_5	−0.343 (0.080)	0.224 (0.262)	−0.129 (0.521)	0.248 (0.211)	−0.544 * (0.003)
K	0.038 (0.849)	−0.250 (0.208)	−0.309 (0.116)	−0.264 (0.184)	0.227 (0.254)
Ca	0.293 (0.138)	0.260 (0.191)	−0.207 (0.299)	0.258 (0.193)	0.255 (0.200)
Mg	0.048 (0.812)	0.252 (0.204)	−0.201 (0.315)	0.241 (0.226)	0.468 * (0.014)
Na	0.111 (0.583)	0.361 (0.064)	0.248 (0.213)	0.366 (0.061)	0.359 (0.066)
CEC	−0.026 (0.899)	0.434 * (0.024)	−0.052 (0.799)	0.459 * (0.016)	−0.350 (0.074)

Correlation coefficients (r) are significantly correlated between the variables compared. Negative values denote negative correlation, and positive values denote positive correlation. Values in parentheses refer to p-values (* $p < 0.05$).

4. Conclusions

The soil bacterial community and diversity of WSG cultivation sites grown in natural conditions in the forest for 13 years had a significant correlation with soil properties such as OM, TN and CEC. Soil pH was recorded to be the most important edaphic factor among the measured soil chemical properties, which drove the abundance of *Acidobacteria* in the studied WSG cultivation sites. This study will enable us to provide a broader idea about the optimum cultivation condition for WSG in natural vegetation condition. In addition, it is believed that more definite information could be provided if a correlation study was conducted on the growth characteristics of WSG and soil bacterial communities according to forest physiognomy and surrounding vegetation along with soil properties.

Author Contributions: K.K., K.S.J. and Y.U., conceptualization of the study; K.K. and H.J.K. designed experiments, performed experiments and analyzed sequencing data; H.J.K., D.H.J. and J.H.H. assisted in soil sampling and experiments; H.J.K. assisted in data analysis and discussion; K.K. wrote the manuscript; K.K. and Y.U. performed critical reviewing and editing. All authors have read and agreed to the published version of the manuscript.

Funding: This work was supported by the research of National Institute of Forest Science (NIFoS) (FP0802-2017-08).

Institutional Review Board Statement: Not applicable.

Informed Consent Statement: Not applicable.

Data Availability Statement: The data and analyses from the current study are available from the corresponding authors upon reasonable request.

Conflicts of Interest: The authors declare no conflict of interest.

References

1. Kim, K.; Huh, J.H.; Um, Y.; Jeon, K.S.; Kim, H.J. The Comparative of Growth Characteristics and Ginsenoside Contents in Wild-simulated Ginseng (*Panax ginseng* C.A. Meyer) on Different Years by Soil Properties of Cultivation Regions. *Korean J. Plant. Resour.* **2020**, *33*, 651–658.
2. Lee, D.S. Weather characteristic and growth of a forest ginseng cultivation site. *J. Korean Soc. For. Sci.* **2011**, *99*, 863–870.
3. Kim, K.; Um, Y.; Jeong, D.H.; Kim, H.J.; Kim, M.J.; Jeon, K.S. Study on the correlation between the soil bacterial community and growth characteristics of wild-simulated ginseng (*Panax ginseng* C.A. Meyer). *Korean J. Environ. Biol.* **2019**, *37*, 380–388. [CrossRef]
4. Falkowski, P.G.; Fenchel, T.; Delong, E.F. The microbial engines that drive Earth's biogeochemical cycles. *Science* **2008**, *320*, 1034–1039. [CrossRef] [PubMed]
5. Prasad, R.; Kumar, M.; Varma, A. Role of PGPR in soil fertility and plant health. In *Plant-Growth-Promoting Rhizobacteria (PGPR) and Medicinal Plant*; Egamberdieva, D., Shrivastava, S., Varma, A., Eds.; Springer: Basel, Switzerland, 2015; pp. 247–260.

6. Busby, P.E.; Soman, C.; Wagner, M.R.; Friesen, M.L.; Kremer, J.; Bennett, A.; Morsy, M.; Eisen, J.A.; Leach, J.E.; Dangl, J.L. Research priorities for harnessing plant microbiomes in sustainable agriculture. *PLoS Biol.* **2017**, *15*, e2001793. [CrossRef]

7. Song, X.; Tao, B.; Guo, J.; Li, J.; Chen, G. Changes in the microbial community structure and soil chemical properties of vertisols under different cropping systems in Northern China. *Front. Environ. Sci.* **2018**, *6*, 132. [CrossRef]

8. Wu, Z.; Liu, Q.; Li, Z.; Cheng, W.; Sun, J.; Guo, Z.; Li, Y.; Zhou, Z.; Meng, D.; Li, H.; et al. Environmental factors shaping the diversity of bacterial communities that promote rice production. *BMC Microbiol.* **2018**, *18*, 51. [CrossRef]

9. Pettersson, M. *Factors Affecting the Rates of Change in Soil Bacterial Communities*; University of Lund: Lund, Sweden, 2004; pp. 17–27.

10. Noll, M.; Wellinger, M. Changes of the soil ecosystem along a receding glacier: Testing the correlation between environmental factors and bacterial community structure. *Soil Biol. Biochem.* **2008**, *40*, 2611–2619. [CrossRef]

11. Gao, P.; Xu, W.; Songtag, P.; Li, X.; Xue, G.; Liu, T.; Sun, W. Correlating microbial community compositions with environmental factors in activated sludge from four full-scale municipal wastewater treatment plants in Shanghai, China. *Environ. Biotechnol.* **2016**, *100*, 4633–4673. [CrossRef]

12. Kim, K.Y.; Samaddar, S.; Chatterjee, P.; Krishnammorthy, R.; Jeon, S.Y.; Sa, T.M. Structural and functional responses of microbial community with respect to salinity levels in a coastal reclamation land. *Appl. Soil Ecol.* **2019**, *137*, 96–105. [CrossRef]

13. Liang, X.; Zhang, Y.; Wommack, K.E.; Wilhelm, S.W.; DeBruyn, J.M.; Sherfy, A.C.; Zhuang, J.; Radosevich, M. Lysogenic reproductive strategies of viral communities vary with soil depth and are correlated with bacterial diversity. *Soil Biol. Biochem.* **2020**, *144*, 107767. [CrossRef]

14. Roy, K.; Ghosh, D.; DeBruyn, J.M.; Dasgupta, T.; Wommack, K.E.; Liang, X.; Wagner, R.E.; Radosevich, M. Temporal dynamics of soil virus and bacterial populations in agricultural and early plant successional soils. *Front. Microbiol.* **2020**, *11*, 1494. [CrossRef] [PubMed]

15. Prasad, R.; Kamal, S.; Sharma, P.K.; Oelmueller, R.; Varma, A. Root endophyte *Piriformospora indica* DSM 11827 alters plants morphology, enhances biomass and antioxidant activity of medicinal plant *Bacopa monniera*. *J. Basic Microbiol.* **2013**, *52*, 1016–1024. [CrossRef] [PubMed]

16. Nguyen, N.L.; Kim, Y.J.; Hoang, V.A.; Subramaniyam, S.; Kang, J.P.; Kang, C.H.; Yang, D.C. Bacterial diversity and community structure in Korean ginseng field soil are shifted by cultivation time. *PLoS ONE* **2016**, *11*, e0155055. [CrossRef] [PubMed]

17. Dong, L.; Xu, J.; Li, Y.; Fang, H.; Niu, W.; Li, X.; Zhang, Y.; Ding, W.; Chen, S. Manipulation of microbial community in the rhizosphere alleviates the replanting issues in *Panax ginseng*. *Soil Biol. Biochem.* **2018**, *125*, 64–74. [CrossRef]

18. Hartmann, M.; Nikalus, P.A.; Zimmermann, S.; Schmutz, S.; Kremer, J.; Abarenkov, K.; Lüscher, P.; Widmer, F.; Frey, B. Resistance and resilience of the forest soil microbiome to logging associated compaction. *ISME J.* **2014**, *8*, 226–244. [CrossRef]

19. Li, H.; Ye, D.; Wang, X.; Settles, M.L.; Wang, J.; Hao, Z.; Zhou, L.; Dong, P.; Jiang, Y.; Ma, Z.S. Soil bacterial communities of different natural forest types in Northeast Chaina. *Plant. Soil* **2014**, *383*, 203–216. [CrossRef]

20. Lee, B.J.; Eo, S.H. Soil bacterial community in red pine forest of Mt. Janggunbong, Bonghwa-gun, Gyeongbuk, Korea, using next generation sequencing. *J. Korean Soc. For. Sci.* **2017**, *106*, 121–129.

21. Kim, K.; Jeong, D.; Kim, H.J.; Jeon, K.; Kim, M.; Um, Y. A study on growth characteristics of wild-simulated ginseng (*Panax ginseng* C. A. Meyer) by direct seeding and transplanting. *Korean J. Plant. Resour.* **2019**, *32*, 160–169.

22. RDA. *Analysis Manual of Comprehensive Examination Laboratory (Soil, Plant, Water and Liquid Manure)*; Rural Development Administration: Suwon, Korea, 2013; pp. 31–53.

23. Canfora, L.; Vendramin, E.; Felici, B.; Tarricone, L.; Florio, A.; Benedetti, A. Vineyard microbiome variations during different fertilization practices revealed by 16S rRNA gene sequencing. *Appl. Soil Ecol.* **2018**, *123*, 71–80. [CrossRef]

24. Schloss, P.D.; Westcott, S.K.; Ryabin, T.; Hall, J.R.; Hartmann, M.; Hollister, E.B.; Lesniewski, R.A.; Oakley, B.B.; Parks, D.H.; Robinson, C.J.; et al. Introducing mothur: Opensource, platform-independent, community-supported software for describing and comparing microbial communities. *Appl. Environ. Microbiol.* **2009**, *75*, 7537–7541. [CrossRef] [PubMed]

25. Schloss, P.D. A high-throughput DNA sequence aligner for microbial ecology studies. *PLoS ONE* **2009**, *4*. [CrossRef] [PubMed]

26. Nurek, T.; Gendek, A.; Roman, K. Forest residues as a renewable source of energy: Elemental composition and physical properties. *BioResources* **2019**, *14*, 6–20. [CrossRef]

27. Schöler, A.; Jacquiod, S.; Vestergaard, G.; Schulz, S.; Schloter, M. Analysis of soil microbial communities based on amplicon sequencing of marker genes. *Biol. Fertil. Soils* **2017**, *53*, 485–489. [CrossRef]

28. Anderson, M.J. *DISTLM v. 5: A FORTRAN Computer Program to Calculate a Distance-Based Multivariate Analysis for a Linear Model*; Department of Statistics, University of Auckland: Auckland, New Zealand, 2004; p. 10.

29. Chung, J.M.; Moon, H.S. Soil Characteristics by the site types around Nari Basin in Ulleung island. *J. Agric. Life Sci.* **2011**, *44*, 44–50.

30. Chamberlain, J.L.; Prisley, S.; Mcguffin, M. Understanding the relationship between American ginseng harvest and hardwood forest inventory and timber harvest to improve co-management of the forest of Eastern United States. *J. Sustain. For.* **2013**, *32*, 605–624. [CrossRef]

31. Woo, S.Y.; Lee, D.S. A study on the growth and environment of *Panax ginseng* in the different forest strands (I). *Korean J. Agric. For. Meteorol.* **2002**, *4*, 65–71.

32. Kim, C.; Choo, G.C.; Cho, H.S.; Lim, J.T. Soil properties of cultivation sites for mountain-cultivated ginseng at local level. *J. Ginseng Res.* **2015**, *39*, 76–80. [CrossRef]

33. Liu, W.W.; Liu, M.C.; Li, W.H.; Zeng, F.S.; Qu, Y. Influence of ginseng cultivation under larch plantations on plant diversity and soil properties in Liaoning Province, Northeast China. *J. Mt. Sci.* **2016**, *13*, 1598–1608. [CrossRef]
34. Binkley, D.; Fisher, R.F. *Ecology and Management of Forest Soils*, 4th ed.; Wiley-Blackwell: Hoboken, NJ, USA, 2013; pp. 39–57.
35. Dong, L.; Li, Y.; Xu, J.; Yang, J.; Wei, G.; Shen, L.; Ding, W.; Chen, S. Biofertilizer regulate the soil microbial community and enhance *Panax ginseng* yields. *Chin. Med.* **2019**, *14*, 20. [CrossRef]
36. Kim, K.Y.; Kim, H.J.; Um, Y.R.; Jeon, K.S. Effect of Soil Properties and Soil Bacterial Community on Early Growth Characteristics of Wild-simulated Ginseng (*Panax ginseng* C.A. Meyer) in Coniferous and Mixed Forest. *Korean J. Medicinal Crop. Sci.* **2020**, *28*, 183–194. [CrossRef]
37. Park, Y.D.; Kwon, T.H.; Eo, S.H. Analysis of soil bacterial community in Ihwaryeong and Yuksimnyeong restoration project sites linking the Ridgeline of Baekdudaegan. *J. Agric. Life Sci.* **2016**, *50*, 117–124. [CrossRef]
38. Rousk, J.; Bååth, E.; Brookes, P.C.; Lauber, C.L.; Lozupone, C.; Caporaso, J.G.; Knight, R.; Fierer, N. Soil bacterial and fungal communities across a pH gradient in an arable soil. *ISME J.* **2010**, *4*, 1340–1351. [CrossRef] [PubMed]
39. Yin, C.T.; Jones, K.L.; Peterson, D.E.; Garrett, K.A.; Hulbert, S.H.; Paulitz, T.C. Members of soil bacterial communities sensitive to tillage and crop rotation. *Soil Biol. Biochem.* **2010**, *42*, 2111–2118. [CrossRef]
40. Zhang, Y.; Cong, J.; Lu, H.; Li, G.; Qu, Y.; Su, X.; Zhou, J.; Li, D. Community structure and elevational diversity patterns of soil *Acidobacteria*. *J. Environ. Sci.* **2010**, *26*, 1717–1724. [CrossRef]
41. Kielak, A.M.; Barreto, C.C.; Kowalchuk, G.A.; van Veen, J.A.; Kuramae, E.E. The Ecology of *Acidobacteria*: Moving beyond Genes and Genomes. *Front. Microbiol.* **2016**, *7*, 744. [CrossRef]
42. Sun, H.; Wang, Q.; Liu, N.; Li, L.; Zhang, C.; Liu, Z.; Zhang, Y. Effects of different leaf litters on the physicochemical properties and bacterial communities in *Panax ginseng*-growing soil. *Appl. Soil Ecol.* **2017**, *111*, 17–24. [CrossRef]
43. Wang, Q.; Sun, H.; Xu, C.; Ma, L.; Li, M.; Shao, C.; Guan, Y.; Liu, N.; Liu, Z.; Zhang, S.; et al. Analysis of rhizosphere bacterial and fungal communities associated with rusty root disease of *Panax ginseng*. *Appl. Soil Ecol.* **2019**, *138*, 245–252. [CrossRef]
44. Mechri, B.; Mariem, F.B.; Baham, M.; Elhadj, S.B.; Hammami, M. Change in soil properties and the soil microbial community following land spreading of olive mill wastewater affects olive trees key physiological parameters and the abundance of arbuscular mycorrhizal fungi. *Soil Biol. Biochem.* **2008**, *40*, 152–161. [CrossRef]
45. Preem, J.K.; Truu, J.; Truu, M.; Mander, Ü.; Oopkaup, K.; Lõhmus, K.; Helmisaari, H.S.; Uri, V.; Zobel, M. Bacterial community structure and its relationship to soil physico-chemical characteristics in alder stands with different management histories. *Ecol. Eng.* **2012**, *49*, 10–17. [CrossRef]
46. Bapiri, A.; Bååth, E.; Rousk, J. Drying-rewetting cycles affect fungal and bacterial growth differently in an arable soil. *Microb. Ecol.* **2010**, *60*, 419–428. [CrossRef] [PubMed]
47. Joa, J.H.; Moon, K.H.; Kim, S.C.; Moon, D.G.; Koh, S.W. Effect of Temperature Condition on Nitrogen Mineralization of Organic Matter and Soil Microbial Community Structure in non-Volcanic Ash Soil. *Korean J. Soil Sci. Fertil.* **2012**, *45*, 377–384. [CrossRef]
48. Altomare, C.; Tringovska, I. Beneficial Soil Microorganisms, an Ecological Alternative for Soil Fertility Management. In *Genetics, Biofuels and Local Farming Systems. Sustainable Agriculture Reviews*; Lichtfouse, E., Ed.; Springer: Dordrecht, The Netherlands; Cham, Switzerland, 2011; pp. 161–214.
49. Lim, S.U. Plant Growth and Nutrients. In *Fertilizer*; Ilsin: Seoul, Korea, 2005; pp. 38–45.
50. Paz-González, A.; Vieira, S.R.; Taboada Castro, M.T. The effect of cultivation on the spatial variability of selected properties of an umbric horizon. *Geoderma* **2000**, *37*, 273–292. [CrossRef]

Article

Interactive Effects of Scion and Rootstock Genotypes on the Root Microbiome of Grapevines (*Vitis* spp. L.)

Stefanie Nicoline Vink [1,*], Francisco Dini-Andreote [1,2], Rebecca Höfle [3], Anna Kicherer [3] and Joana Falcão Salles [1,*]

1 Groningen Institute for Evolutionary Life Sciences, University of Groningen, 9747AG Groningen, The Netherlands

2 Department of Plant Science & Huck Institutes of the Life Sciences, The Pennsylvania State University, University Park, State College, PA 16802, USA; fjd5141@psu.edu

3 Institute for Grapevine Breeding Geilweilerhof, Julius Kühn-Institut, 76833 Siebeldingen, Germany; rebecca.hoefle@julius-kuehn.de (R.H.); anna.kicherer@julius-kuehn.de (A.K.)

* Correspondence: vinksn@gmail.com (S.N.V.); j.falcao.salles@rug.nl (J.F.S.); Tel.: +31-50-363-2162 (J.F.S.)

Abstract: Diversity and community structure of soil microorganisms are increasingly recognized as important contributors to sustainable agriculture and plant health. In viticulture, grapevine scion cultivars are grafted onto rootstocks to reduce the incidence of the grapevine pest phylloxera. However, it is unknown to what extent this practice influences root-associated microbial communities. A field survey of bacteria in soil surrounding the roots (rhizosphere) of 4 cultivars × 4 rootstock combinations was conducted to determine whether rootstock and cultivar genotypes are important drivers of rhizosphere community diversity and composition. Differences in α-diversity was highly dependent on rootstock–cultivar combinations, while bacterial community structure primarily clustered according to cultivar differences, followed by differences in rootstocks. Twenty-four bacterial indicator genera were significantly more abundant in one or more cultivars, while only thirteen were found to be specifically associated with one or more rootstock genotypes, but there was little overlap between cultivar and rootstock indicator genera. Bacterial diversity in grafted grapevines was affected by both cultivar and rootstock identity, but this effect was dependent on which diversity measure was being examined (i.e., α- or β-diversity) and specific rootstock–cultivar combinations. These findings could have functional implications, for instance, if specific combinations varied in their ability to attract beneficial microbial taxa which can control pathogens and/or assist plant performance.

Keywords: agricultural practices; cultivar; grafting; interaction rootstock scion; plant performance; rhizosphere bacteria; taxonomic indicators; viticulture

Citation: Vink, S.N.; Dini-Andreote, F.; Höfle, R.; Kicherer, A.; Salles, J.F. Interactive Effects of Scion and Rootstock Genotypes on the Root Microbiome of Grapevines (*Vitis* spp. L.). *Appl. Sci.* **2021**, *11*, 1615. https://doi.org/10.3390/app11041615

Academic Editor: Tongmin Sa

Received: 17 December 2020

Accepted: 4 February 2021

Published: 10 February 2021

Publisher's Note: MDPI stays neutral with regard to jurisdictional claims in published maps and institutional affiliations.

1. Introduction

Soil microbial communities perform a wide range of ecosystem processes and functions and are increasingly recognized as vital components of a healthy agroecosystem. For instance, bacteria play pivotal roles in the biogeochemical cycling of nutrients, and influence plant productivity and health through the action of specific plant growth-promoting (PGPB) and biocontrol bacterial species, and/or negatively, through the actions of plant pathogens [1,2]. Rhizosphere bacterial diversity is known to affect plant health, with communities with a higher diversity generally better able to withstand invasion of pathogens and possessing higher amounts of PGPBs [3–5]. The higher biodiversity often relates to increasing levels of ecosystem functions and services [6,7]. In addition, plant functional traits, such as abiotic and biotic tolerance (e.g., salinity, drought, diseases) and nutrient uptake, are to varying degrees directly influenced by the root-associated microbial communities [8–10].

Conversely, soil communities are also affected by plant characteristics, primarily through the production of root exudates. For instance, up to 40% of the photosynthates produced by a plant can be actively released by the roots [11], and the quality and quantity of these carbon compounds can vary between and within plant species [12–14], potentially leading to highly specific relationships between plants and microorganisms [15,16], such as those between rhizobia and legumes [17] or with plant pathogens [18,19]. Bacterial communities are also known to differ according to plant genotypes and hosts [12,15,20,21], although this effect is by no means ubiquitous, and plant genotypic differences do not always lead to significant differences in microbiomes in the rhizosphere [22].

Viticulture is one of the world's main horticultural practices. In 2018, approximately 77.8 million tons of grapes were produced globally, primarily for wine production [23]. Grafting is common practice in viticulture to reduce the incidence of the grapevine pest phylloxera; European *Vitis vinifera* L. scions (i.e., the upper plant) merged with North American *Vitis* sp. hybrid rootstocks [24]. This results in plants with stress- and disease-resistance combined with desirable agronomic characteristics for grape production and allows breeders to select specific traits independently for rootstocks and cultivars.

There is a considerable body of research on the effect of rootstock on scion physiology and physical properties in viticulture, for instance, by showing improved drought tolerance [25,26], and changes in stomatal conductance and transpiration [27]. Less research has been conducted on the effects of the scion on the rootstock, and, to date, this effect has been examined only on physical rootstock properties such as root biomass and length [28]. The interaction between rootstocks and scions also remains understudied, but research has shown that it can affect yield and quality of grapes [29], and root behavior and plant growth [30]. With grafting, scions and rootstocks each maintain their own genetic identity [24], and in other plant species this has led to changes in bacterial diversity [31] or community structure [32].

Despite the potential influence of grapevine genotypic variation on bacterial community structure and functioning in the rhizosphere, and the importance of the root microbiome for plant development and health, little is known about how rootstock and scion, and importantly, their interaction, exert an influence on soil bacterial communities in grapevines. Therefore, a survey was conducted with the aim of elucidating the effect of different rootstock and grafted scion combinations on soil bacterial communities associated with the rhizosphere of grapevines.

2. Materials and Methods

2.1. Experimental Design and Soil Sampling

In October 2016, a field survey of bacteria from the rhizosphere of different scion-rootstock combinations was conducted. Grapevines were planted 11 years prior to the experiment in a vineyard at the Institute for Grapevine Breeding, Julius Kühn-Institut in Siebeldingen, Germany. The vineyard was located in an area with sandy loam soil (type pararedzina). In 2016, the area received approximately 625 mm rainfall.

Each of four scion cultivars (Calandro, Reberger, Felicia, and Villaris; from now on referred to as cultivars) were grafted onto each of four rootstock types (125 AA, 5 BB, Binova, and SO 4). All cultivars belong to the *Vitis vinifera* species; Calandro and Reberger are red grape varieties, and Felicia and Villaris are white grapes. The rootstocks were all *Vitis berlandieri* and *Vitis riparia* crosses with low genetic variation [33,34]. Grafted grapevines were placed in adjacent rows which were all subjected to cover cropping: every second row with a vineyard-specific mixture of various legumes and other forbs (Wolff-mix) and tilled via disc harrow in spring each year, and the remaining rows with grass and mulched three times between April and August, depending on the weather and growing conditions. Fertilization was conducted yearly with Entec 26 (26% N, 13% S) at a rate of 39 kg N/ha. For the duration of the experiment no irrigation took place.

Rhizosphere soils were collected in a single day, from three randomly selected plants from each cultivar x rootstock combination, yielding a total of 48 samples. Samples were collected by removing roots with attached soil in the field using sterile metal forceps, cleaned with 70% ethanol between each sample to avoid cross-contamination. Samples were transported on ice and processed in the laboratory within 24 h. Sampled roots were shaken to remove loosely adhering soil, and rhizosphere soil was collected by removing the remaining firmly attached soil particles using a sterile disposable brush. Samples were stored at $-20°$ C for further DNA extraction.

2.2. DNA Extraction and Sequencing

DNA was extracted using the PowerSoil DNA extraction kit (Qiagen, Hilden, Germany) following the manufacturer's instructions, and 0.5 g of rhizosphere soil per sample, with an addition of 50 mg of sterile glass beads (ø 0.1 mm, BioSpec Products, Bartlesville, WA, USA) to improve extraction yield. DNA quality and quantity were determined using a NanoDrop 2000 Spectrophotometer (Thermo Fisher scientific, Walthan, MA, USA). Amplicon library preparation and sequencing based on the V4 region of the bacterial 16S rRNA gene was conducted at the Argonne National Laboratory Sequencing Facility (Lemont, IL, USA) using the Illumina MiSeq platform, following the Earth Microbiome Project protocol [35], yielding paired-end reads of 150 bp in length.

2.3. Sequence Quality Filtering and Data Analysis

Reads were imported into QIIME2 version 2018.2 [36], demultiplexed and primers were removed using the EMP procedure. Sequences were filtered, dereplicated and denoised, chimeric sequences were removed, and paired using the DADA2 plugin [37] in QIIME2, resulting in amplicon sequence variants (ASVs) of 253 bp in length. ASVs were aligned using MAFFT [38] and used to generate a mid-point rooted phylogenetic tree using FastTree [39]. Taxonomic assignment was performed using the RDP classifier version 2.10.1 [40].

All statistical analyses were conducted in R version 3.6.2 [41]. The ASV table, taxonomy table, phylogenetic tree and associated metadata were imported into the package Phyloseq version 1.30.0 [42], and any non-bacterial ASVs and those taxonomically affiliated to chloroplast and mitochondrion were removed. The resulting table was rarefied to 15,704 sequences per sample to account for any differences in sequencing depth.

Observed ASV richness, Shannon and Simpson diversity indices were calculated using Phyloseq and Faith's phylogentic diversity (PD) in the packages picante version 1.8 [43] and btools version 0.0.1 [44]. Significant differences in α-diversity measures (i.e., variation of species within a sample) between cultivars, rootstocks and their interaction were tested using ANOVA on Aligned Rank Transformed data (i.e., a non-parametric ANOVA) using the package ARTool version 0.10.7 [45]. This method was used because the α-diversity measures did not meet the assumption of normally distributed and/or homogeneity of variance for a parametric ANOVA. As PD showed the strongest effect of treatment, a post hoc analysis was performed for the rootstock $\times$ cultivar interaction term using estimated marginal means with the package emmeans version 1.4.5 [46].

Bacterial community structure (β-diversity, i.e., variation in species between samples) was assessed using principal coordinate analysis (PCoA) based on Bray-Curtis distances, which calculates dissimilarity between samples using relative sequence abundances. To determine potential significant effects of rootstock, cultivar and their interaction, PERMANOVA with 999 permutations was used with the package vegan version 2.5.4 [47]. Bacterial taxonomic composition was evaluated by merging taxa to the phylum level (and summing their abundances), and removing those with an abundance <20. Data were plotted based on relative abundances per rootstock and cultivar combination. To detect bacterial taxa with an affinity towards specific (combinations of) cultivars or rootstocks, indicator species analysis using the package indicspecies [48] was performed using abundance data of taxa merged at the genus level. The analysis was carried out separately for rootstock and cultivar samples and resulted in a list of species associated with individual, or groups of, cultivars or rootstocks, based on calculations that take into account the strength of association and the statistical significance of the relationship between species abundances and treatment groups.

3. Results

3.1. Bacterial α-Diversity across Different Cultivar-Rootstock Combinations

Rootstock identity was found to exert a significant influence on rhizosphere bacteria for all α-diversity measures, while cultivar identity only significantly affected Simpson and Faith's PD indices (Table 1). However, all α-diversity measures showed a significant interaction between cultivars and rootstocks (Table 1) indicating that the observed differences between rootstocks varied in strength according to cultivar identity.

Table 1. ANOVA results of α-diversity measures per rootstock and cultivar identity and the interaction between the two factors. Data were aligned rank transformed prior to statistical analysis. Significance levels are as follows: *** $p < 0.001$, ** $p < 0.01$, * $p < 0.05$, ns = not significant.

	Richness	Shannon	Simpson	Faith's PD
Rootstock	$F_{3,32} = 4.6$ **	$F_{3,32} = 5.2$ **	$F_{3,32} = 3.2$ *	$F_{3,32} = 7.3$ ***
Cultivar	$F_{3,32} = 0.9$ ns	$F_{3,32} = 2.6$ ns	$F_{3,32} = 4.5$ **	$F_{3,32} = 7.1$ ***
Rootstock:Cultivars	$F_{9,32} = 3.9$ **	$F_{9,32} = 4.0$ **	$F_{9,32} = 3.0$ **	$F_{9,32} = 4.8$ ***

Of all α-diversity measures, Faith's PD showed the largest difference for both rootstocks (Figure 1a) and cultivars (Figure 1b) as main effects, in addition to their interaction (Figure 1c). This interaction effect was also observed for the other diversity measures (Figure S1). Post hoc analysis on Faith's PD indicated that the most distinct rootstock–cultivar combinations were 5 BB-Reberger, 125 AA-Calandro and 125 AA-Vilaris. However, values of Faith's PD were quite similar for the majority of cultivar x rootstock combinations.

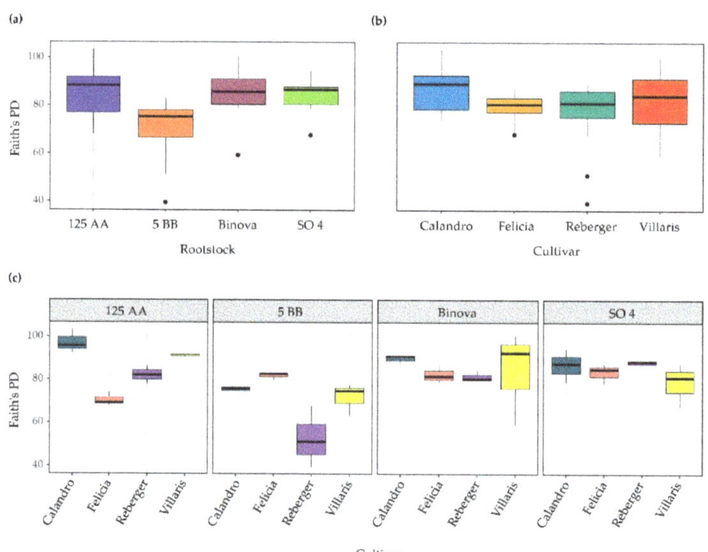

Figure 1. Faith's phylogenetic diversity of bacterial rhizosphere communities per (**a**) cultivar, (**b**) rootstock and (**c**) the interaction between rootstock and cultivar.

3.2. Bacterial β-Diversity across Different Cultivar-Rootstock Combinations

The main determinant of bacterial community structure in the rhizosphere was cultivar identity (PERMANOVA: $R^2 = 0.14$, $p < 0.001$), with samples from Calandro and Villaris plants harboring a different community to those of Felicia and Reberger (Figure 2a).

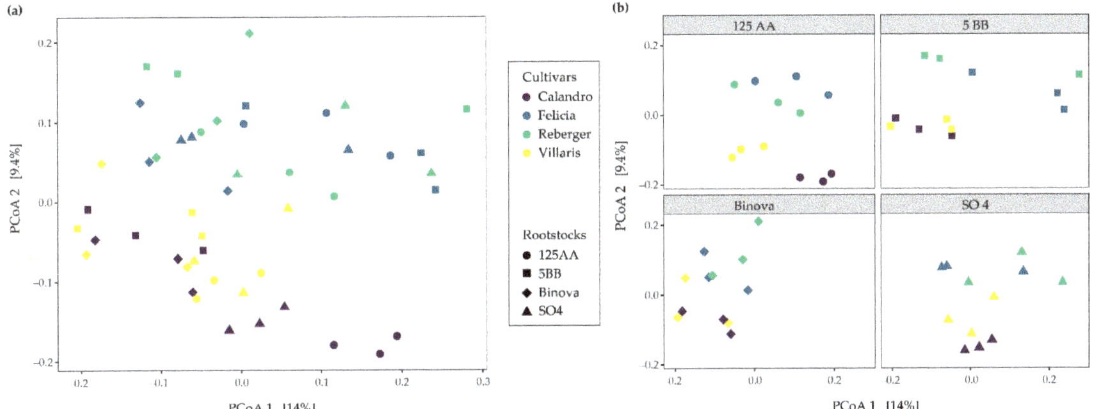

Figure 2. PCoA of rhizosphere soil bacterial communities based on the Bray-Curtis dissimilarities. The figure displays (**a**) collection of all samples or (**b**) plots separated according to each rootstock.

Although this separation was independent of rootstock identity, i.e., there was no significant interaction effect (PERMANOVA: $R^2 = 0.18$, $p = 0.65$), in particular the combination 125 AA-Calandro and 5BB-Calandro harbored a distinct bacterial community from the other cultivar–rootstock combinations (Figure 2a). Rootstock identity was found to exert a comparatively minor significant effect on differences in bacterial community structure (PERMANOVA: $R^2 = 0.10$, $p < 0.001$). However, such differences were less visible than that imposed by differences in cultivar identity within each rootstock.

3.3. Taxonomic Level Analysis

Taxonomic composition at the phylum level did not vary greatly between the different cultivar–rootstock combinations (Figure 3). For all combinations, the phyla of *Proteobacteria*, *Acidobacteria*, *Actinobacteria*, *Bacteroidetes* and *Firmicutes* had the highest relative abundance, and while the abundances of the less dominant phyla varied per rootstock–cultivar combination, here there were also no large differences observed.

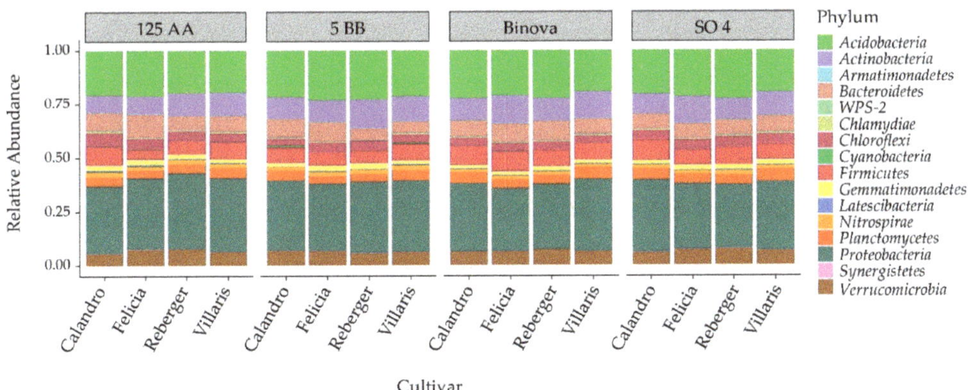

Figure 3. Taxonomic composition of rhizosphere bacterial communities per rootstock and cultivar. Bars show the relative abundance of bacterial taxa merged at the phylum level. Phyla whose abundance was <20 were not shown for clarity of data presentation.

3.4. Bacterial Indicator Species Analysis

Indicator species analysis was conducted separately for cultivars and rootstocks at the genus level. Twenty-four bacterial genera were associated with one or more cultivars (Figure 4a), while only thirteen were found to be specifically associated with the rootstocks (Figure 4b), and only one genus (*Elioraea*) was associated with both rootstock and cultivars. The number of genera that associated with a single cultivar or rootstock were low-*Marmoricola* was specifically associated with one cultivar (Reberger) while the genera *Escherichia/Shigella* and *Enterobacter* were exclusively associated with the rootstock 125 AA. Instead, the vast majority of bacterial genera were indicative of multiple groups, with the majority associated to 3 (out of the 4) groups for both cultivars (Figure 4a) and rootstocks (Figure 4b).

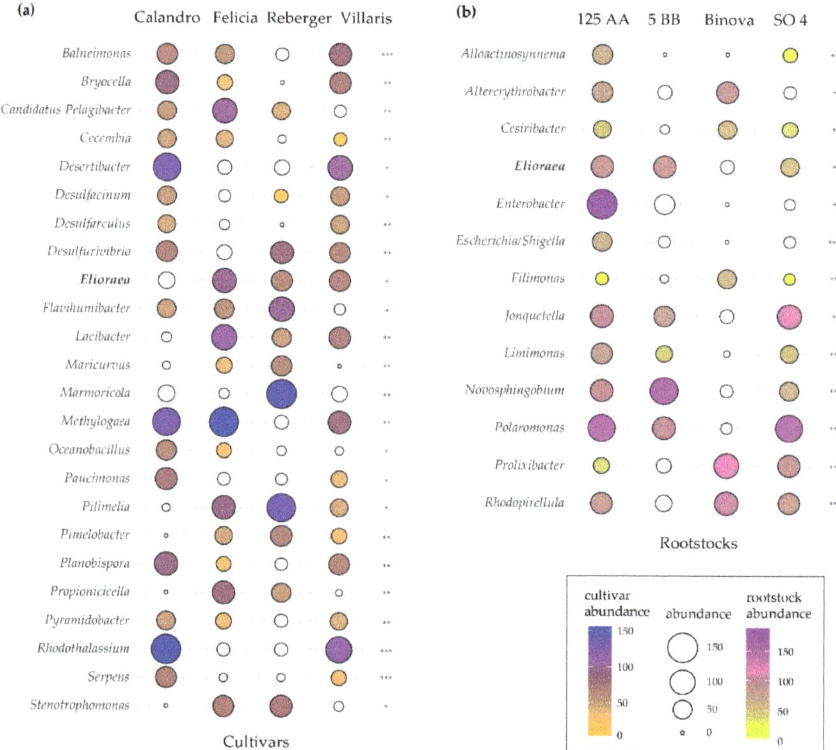

Figure 4. Indicator species analysis of bacterial genera per cultivars (**a**) or rootstocks (**b**). Color and bubble size relate to sequence abundance per genus, bubbles without color indicate that the genus was not an indicator for that particular cultivar or rootstock. *** $p < 0.001$, ** $p < 0.01$, * $p < 0.05$. Genera that are indicator species in both datasets are shown in bold.

The genus *Elioraea*, which was an indicator species in both rootstocks and cultivars, is relatively unknown, with few known members; the genus is thought to belong to the aerobic anoxygenic photosynthetic bacteria [49], which are obligate aerobes that can capture energy from light through photosynthesis [50]. Species in the genus *Marmoricola* are all mesophilic, non-pathogenic and often found in soil environments [51]. The genera *Escherichia/Shigella* and *Enterobacter* all belong to the Enterobacteriaceae, and have members which are known human, animal and plant pathogens, as well as plant-associated growth-promoting species, and lignin degraders [52]. The genera *Desulfacinum*, *Desulfarculus* and *Desulurivibrio*, which were indicator species for all but the cultivar Felicia, are known to be anaerobic sulfate reducers [53].

4. Discussion

The aim of this study was to determine whether rootstock and cultivar identity are important drivers of the structure and composition of rhizosphere communities in grapevines. Based on the results from experiments on other plant species [31,32] and the limited knowledge in grapevines [9,54], the expectation was that there would be an effect of grafting on rhizosphere microbial communities of different cultivar–rootstock combinations. For many of the measures that were examined, such as α-diversity and taxonomic composition, highly specific cultivar–rootstock interaction effects were observed, but on the whole, these effects occurred for a few specific rootstock–cultivar combinations.

However, a clear differentiation in bacterial community structure (β-diversity) was observed, and in particular a clear distinction was observed between two groups of cultivars; Calandro and Villaris on the one hand and Felicia and Reberger on the other. There is clear genetic component to plant root exudate composition [22,55,56] and this can strongly affect rhizosphere bacterial communities in *Vitis* sp. [57]. The two red cultivars (Calandro and Reberger) were more closely related to each other (sharing one parent) than to the two white grapevine cultivars, which were also closely related (sharing both parents) [34]. Therefore, the expectation was that there would be a differentiation along red and white cultivars, but this was not the case. Due to the genetic basis of root exudate composition, examination of root exudate quality, and its effect on bacterial communities, might offer more insight into the causes of the observed effect.

Less difference was observed between the different rootstock genotypes. While the cultivars consisted of genotypes from very different parents, the rootstocks all had a similar lineage [33] and therefore were likely to be more similar genetically (and phenotypically) than the cultivars. Previous studies have shown clear differences in bacterial communities between different rootstock genotypes grafted with a single scion genotype [54]. However, this can be context dependent, and rhizosphere bacterial communities have been shown to differ between rootstocks in one vineyard and not in another [58]. In addition, Berlanas et al. [58] found no difference in rhizosphere microbiomes between different 7-year-old grapevine rootstock genotypes, but that mature (25 years old) rootstock genotypes were associated with different rhizosphere microbiomes. In this context, given that the grapevines used in the current study were 11 years old, the specific effect of rootstock genotype might not have yet developed. Many aspects of both host and environment are responsible for the quality and quantity of root exudates [55], and hence in shaping rhizosphere bacterial communities, and it seems likely that a combination of factors played a role in shaping bacterial communities in the current experiment.

The α-diversity measures were determined to a large degree by the interaction between rootstock and cultivar. Although most combinations did not differ significantly, there were some notable exceptions, such as the combinations 5 BB-Reberger and 125 AA-Felicia with low, and 125 AA-Calandro with high α-diversity values. This variation could have functional implications as microbial species richness has been linked to resistance to invasion of pathogens [3,4]. High species richness, particularly combined with a more even distribution of those species, is often paired with improved ecosystem services and stability, and enhanced plant performance due to beneficial traits [59–61]. In the current study, the strongest response was observed with Faiths Phylogenetic Diversity, implying a potentially higher diversity in terms of functions [62] and resistance to invasion [63].

There were no large differences between any of the rootstock–cultivar combinations at the phylum level in the current study. The rhizosphere soils were dominated by the phyla Proteobacteria, Acidobacteria, Actinobacteria, Bacteriodetes and Firmicutes, which is in line with soils found across the globe [6], as well as in vineyards [64,65]. However, functional effects of the microbiome on plants are more likely to occur at lower taxonomic levels, and individual taxa can also make a large impact, such as with biocontrol, PGP and pathogenic bacteria [1,2]. Indicator species analysis showed that, on the whole, genera were differentially abundant in more than one cultivar or rootstock genotype, and at least some of these genera are known to differ in terms of their functionality. Five bacterial genera were in relatively higher abundance in the rhizosphere of the cultivars Calandro and Villaris, which grouped together in terms of bacterial community structure, than in the other two cultivars. For this cultivar group, two genera *Desertibacter* and *Rhodothalassium* occurred in relatively high abundance. Both of these genera belong to families [66,67] with members known to be able to benefit plant growth [68]. Three genera were specifically associated with the other β-diversity cultivar group (Felicia and Reberger), but here, only one genus (*Stenotrophomonas*) was in high abundance. Species in the *Stenotrophomonas* are known to be closely associated with plants, and also have known plant beneficial effects [69]. Although speculative, it seems that both groups of cultivars are able to recruit (i.e., attract and select) different, potentially beneficial, genera.

5. Conclusions

The results from this study reveal for this first time that bacterial diversity in grafted grapevines is affected by both cultivar and rootstock identity, but that this effect is dependent on which diversity measure is being examined (i.e., α- or β-diversity) and specific rootstock–cultivar combinations. This finding is of particular importance since it could affect the ability of specific combinations to chemically attract, via root exudates, specific bacteria that could affect key functions associated with plant protection, performance, and productivity [70], as well as specifically to viticulture because it can affect colonization in other plant organs such as grapes [65], potentially impacting wine quality [71]. A first step toward further understanding the nature and cause of the observed rootstock x scion interactive effects on soil bacterial communities in this study would be to unravel the link between grapevine genomes with their root exudation patterns, as well as to their rhizosphere bacterial communities.

Supplementary Materials: The following are available online at https://www.mdpi.com/2076-3417/11/4/1615/s1, Figure S1: observed richness, and Shannon and Simpson diversity indices of bacterial rhizosphere communities per rootstock and cultivar interaction.

Author Contributions: Conceptualization, F.D.-A. and J.F.S.; Data curation, S.N.V.; Formal analysis, S.N.V.; Funding acquisition, J.F.S.; Investigation, F.D.-A., R.H. and A.K.; Methodology, F.D.-A.; Project administration, S.N.V., F.D.-A. and J.F.S.; Resources, R.H. and A.K.; Software, S.N.V.; Supervision, J.F.S.; Visualization, S.N.V.; Writing—original draft, S.N.V.; Writing—review and editing, S.N.V., F.D.-A., R.H., A.K. and J.F.S. All authors have read and agreed to the published version of the manuscript.

Funding: This research was co-financed by the European Union (FACCE SURPLUS ERA-NET grant 652615) and the Netherlands Organization for Scientific Research (NOW—grant number ALW.FACCE.1), as part of the international collaborative 'Vitismart' project.

Institutional Review Board Statement: Not applicable.

Informed Consent Statement: Not applicable.

Data Availability Statement: The sequencing reads from this study are openly available in the European Nucleotide Archive under accession nr. PRJEB42324. This data can be found at https://www.ebi.ac.uk/ena/browser/view/PRJEB42324 (accessed on 20 January 2021).

Acknowledgments: The authors would like to thank Jolanda Brons for assistance with sampling and Jan Veldsink for assistance with molecular methods. We would also like to thank the Center for Information Technology of the University of Groningen for their support and for providing access to the Peregrine high-performance computing cluster.

Conflicts of Interest: The authors declare no conflict of interest. The funders had no role in the design of the study; in the collection, analyses, or interpretation of data; in the writing of the manuscript, or in the decision to publish the results.

References

1. Lugtenberg, B.; Kamilova, F. Plant-Growth-Promoting Rhizobacteria. *Annu. Rev. Microbiol.* **2009**, *63*, 541–556. [CrossRef]
2. Berendsen, R.L.; Pieterse, C.M.J.; Bakker, P.A.H.M. The rhizosphere microbiome and plant health. *Trends Plant Sci.* **2012**, *17*, 478–486. [CrossRef] [PubMed]

3. van Elsas, J.D.; Chiurazzi, M.; Mallon, C.A.; Elhottova, D.; Kristufek, V.; Salles, J.F. Microbial diversity determines the invasion of soil by a bacterial pathogen. *Proc. Natl. Acad. Sci. USA* **2012**, *109*, 1159–1164. [CrossRef]

4. Mallon, C.A.; van Elsas, J.D.; Salles, J.F. Microbial Invasions: The Process, Patterns, and Mechanisms. *Trends Microbiol.* **2015**, *23*, 719–729. [CrossRef]

5. Kurm, V.; van der Putten, W.H.; Pineda, A.; Hol, W.H.G. Soil microbial species loss affects plant biomass and survival of an introduced bacterial strain, but not inducible plant defences. *Ann. Bot.* **2018**, *121*, 311–319. [CrossRef]

6. Delgado-Baquerizo, M.; Maestre, F.T.; Reich, P.B.; Jeffries, T.C.; Gaitan, J.J.; Encinar, D.; Berdugo, M.; Campbell, C.D.; Singh, B.K. Microbial diversity drives multifunctionality in terrestrial ecosystems. *Nat. Commun.* **2016**, *7*, 10541. [CrossRef] [PubMed]

7. Wagg, C.; Schlaeppi, K.; Banerjee, S.; Kuramae, E.E.; van der Heijden, M.G.A. Fungal-bacterial diversity and microbiome complexity predict ecosystem functioning. *Nat. Commun.* **2019**, *10*, 4841. [CrossRef]

8. Goh, C.-H.; Veliz Vallejos, D.F.; Nicotra, A.B.; Mathesius, U. The Impact of Beneficial Plant-Associated Microbes on Plant Phenotypic Plasticity. *J. Chem. Ecol.* **2013**, *39*, 826–839. [CrossRef]

9. Rolli, E.; Marasco, R.; Vigani, G.; Ettoumi, B.; Mapelli, F.; Deangelis, M.L.; Gandolfi, C.; Casati, E.; Previtali, F.; Gerbino, R.; et al. Improved plant resistance to drought is promoted by the root-associated microbiome as a water stress-dependent trait: Root bacteria protect plants from drought. *Environ. Microbiol.* **2015**, *17*, 316–331. [CrossRef] [PubMed]

10. Friesen, M.L.; Porter, S.S.; Stark, S.C.; von Wettberg, E.J.; Sachs, J.L.; Martinez-Romero, E. Microbially Mediated Plant Functional Traits. *Annu. Rev. Ecol. Evol. Syst.* **2011**, *42*, 23–46. [CrossRef]

11. Bais, H.P.; Weir, T.L.; Perry, L.G.; Gilroy, S.; Vivanco, J.M. The role of root exudates in rhizosphere interactions with plants and other organisms. *Annu. Rev. Plant Biol.* **2006**, *57*, 233–266. [CrossRef]

12. Berg, G.; Smalla, K. Plant species and soil type cooperatively shape the structure and function of microbial communities in the rhizosphere: Plant species, soil type and rhizosphere communities. *FEMS Microbiol. Ecol.* **2009**, *68*, 1–13. [CrossRef]

13. Doornbos, R.F.; van Loon, L.C.; Bakker, P.A.H.M. Impact of root exudates and plant defense signaling on bacterial communities in the rhizosphere. A review. *Agron. Sustain. Dev.* **2012**, *32*, 227–243. [CrossRef]

14. İnceoğlu, Ö.; Al-Soud, W.A.; Salles, J.F.; Semenov, A.V.; van Elsas, J.D. Comparative Analysis of Bacterial Communities in a Potato Field as Determined by Pyrosequencing. *PLoS ONE* **2011**, *6*, e23321. [CrossRef]

15. Jiang, Y.; Li, S.; Li, R.; Zhang, J.; Liu, Y.; Lv, L.; Zhu, H.; Wu, W.; Li, W. Plant cultivars imprint the rhizosphere bacterial community composition and association networks. *Soil Biol. Biochem.* **2017**, *109*, 145–155. [CrossRef]

16. Oger, P.M.; Mansouri, H.; Nesme, X.; Dessaux, Y. Engineering Root Exudation of Lotus toward the Production of Two Novel Carbon Compounds Leads to the Selection of Distinct Microbial Populations in the Rhizosphere. *Microb. Ecol.* **2004**, *47*, 96–103. [CrossRef]

17. Wang, D.; Yang, S.; Tang, F.; Zhu, H. Symbiosis specificity in the legume—Rhizobial mutualism: Host specificity in root nodule symbiosis. *Cell. Microbiol.* **2012**, *14*, 334–342. [CrossRef] [PubMed]

18. Anderson, A.R. Host Specificity in the Genus *Agrobacterium*. *Phytopathology* **1979**, *69*, 320. [CrossRef]

19. Jacques, M.-A.; Arlat, M.; Boulanger, A.; Boureau, T.; Carrère, S.; Cesbron, S.; Chen, N.W.G.; Cociancich, S.; Darrasse, A.; Denancé, N.; et al. Using Ecology, Physiology, and Genomics to Understand Host Specificity in *Xanthomonas*. *Annu. Rev. Phytopathol.* **2016**, *54*, 163–187. [CrossRef] [PubMed]

20. Liu, F.; Hewezi, T.; Lebeis, S.L.; Pantalone, V.; Grewal, P.S.; Staton, M.E. Soil indigenous microbiome and plant genotypes cooperatively modify soybean rhizosphere microbiome assembly. *BMC Microbiol.* **2019**, *19*, 201. [CrossRef] [PubMed]

21. Mendes, L.W.; Raaijmakers, J.M.; de Hollander, M.; Mendes, R.; Tsai, S.M. Influence of resistance breeding in common bean on rhizosphere microbiome composition and function. *ISME J.* **2018**, *12*, 212–224. [CrossRef]

22. Wagner, M.R.; Lundberg, D.S.; del Rio, T.G.; Tringe, S.G.; Dangl, J.L.; Mitchell-Olds, T. Host genotype and age shape the leaf and root microbiomes of a wild perennial plant. *Nat. Commun.* **2016**, *7*, 12151. [CrossRef] [PubMed]

23. International Organisation of Vine and Wine. *2019 Statistical Report on World Vitiviniculture*; OIV: Paris, France, 2019.

24. Mudge, K.; Janick, J.; Scofield, S.; Goldschmidt, E.E. A History of Grafting. In *Horticultural Reviews*; Janick, J., Ed.; John Wiley & Sons, Inc.: Hoboken, NJ, USA, 2009; pp. 437–493. ISBN 978-0-470-59377-6.

25. Alsina, M.M.; Smart, D.R.; Bauerle, T.; de Herralde, F.; Biel, C.; Stockert, C.; Negron, C.; Save, R. Seasonal changes of whole root system conductance by a drought-tolerant grape root system. *J. Exp. Bot.* **2011**, *62*, 99–109. [CrossRef]

26. Serra, I.; Strever, A.; Myburgh, P.A.; Deloire, A. Review: The interaction between rootstocks and cultivars (*Vitis vinifera* L.) to enhance drought tolerance in grapevine: Rootstocks to enhance drought tolerance in grapevine. *Aust. J. Grape Wine Res.* **2014**, *20*, 1–14. [CrossRef]

27. Peccoux, A.; Loveys, B.; Zhu, J.; Gambetta, G.A.; Delrot, S.; Vivin, P.; Schultz, H.R.; Ollat, N.; Dai, Z. Dissecting the rootstock control of scion transpiration using model-assisted analyses in grapevine. *Tree Physiol.* **2018**, *38*, 1026–1040. [CrossRef] [PubMed]

28. Tandonnet, J.-P.; Cookson, S.J.; Vivin, P.; Ollat, N. Scion genotype controls biomass allocation and root development in grafted grapevine: Scion/rootstock interactions in grapevine. *Aust. J. Grape Wine Res.* **2009**, *16*, 290–300. [CrossRef]

29. Clingeleffer, P.; Morales, N.; Davis, H.; Smith, H. The significance of scion × rootstock interactions. *OENO ONE* **2019**, *53*. [CrossRef]

30. Ferlito, F.; Distefano, G.; Gentile, A.; Allegra, M.; Lakso, A.N.; Nicolosi, E. Scion–rootstock interactions influence the growth and behaviour of the grapevine root system in a heavy clay soil. *Aust. J. Grape Wine Res.* **2020**, *26*, 68–78. [CrossRef]

31. Ling, N.; Song, Y.; Raza, W.; Huang, Q.; Guo, S.; Shen, Q. The response of root-associated bacterial community to the grafting of watermelon. *Plant Soil* **2015**, *391*, 253–264. [CrossRef]

32. Poudel, R.; Jumpponen, A.; Kennelly, M.M.; Rivard, C.L.; Gomez-Montano, L.; Garrett, K.A. Rootstocks Shape the Rhizobiome: Rhizosphere and Endosphere Bacterial Communities in the Grafted Tomato System. *Appl. Environ. Microbiol.* **2018**, *85*, e01765-18. [CrossRef] [PubMed]

33. Ruehl, E.; Schmid, J.; Eibach, R.; Töpfer, R. Grapevine breeding programmes in Germany. In *Grapevine Breeding Programs for the Wine Industry*; Elsevier: Amsterdam, The Netherlands, 2015; pp. 77–101, ISBN 978-1-78242-075-0.

34. Maul, E.; Toepfer, R. Vitis International Variety Catalogue. 2020. Available online: www.vivc.de (accessed on 29 November 2020).

35. Caporaso, J.G.; Ackermann, G.; Apprill, A.; Bauer, M.; Berg, D.; Betley, J.; Fierer, N.; Fraser, L.; Fuhrman, A.J.; Gilbert, A.J.; et al. EMP 16S Illumina Amplicon Protocol. Available online: https://www.protocols.io/view/emp-16s-illumina-amplicon-protocol-nuudeww (accessed on 3 February 2018).

36. Bolyen, E.; Rideout, J.R.; Dillon, M.R.; Bokulich, N.A.; Abnet, C.C.; Al-Ghalith, G.A.; Alexander, H.; Alm, E.J.; Arumugam, M.; Asnicar, F.; et al. Reproducible, interactive, scalable and extensible microbiome data science using QIIME 2. *Nat. Biotechnol.* **2019**, *37*, 852–857. [CrossRef]

37. Callahan, B.J.; McMurdie, P.J.; Rosen, M.J.; Han, A.W.; Johnson, A.J.A.; Holmes, S.P. DADA2: High-resolution sample inference from Illumina amplicon data. *Nat. Methods* **2016**, *13*, 581–583. [CrossRef]

38. Katoh, K.; Standley, D.M. MAFFT Multiple Sequence Alignment Software Version 7: Improvements in Performance and Usability. *Mol. Biol. Evol.* **2013**, *30*, 772–780. [CrossRef] [PubMed]

39. Price, M.N.; Dehal, P.S.; Arkin, A.P. FastTree 2—Approximately Maximum-Likelihood Trees for Large Alignments. *PLoS ONE* **2010**, *5*, e9490. [CrossRef]

40. Wang, Q.; Garrity, G.M.; Tiedje, J.M.; Cole, J.R. Naïve Bayesian Classifier for Rapid Assignment of rRNA Sequences into the New Bacterial Taxonomy. *AEM* **2007**, *73*, 5261–5267. [CrossRef] [PubMed]

41. R Core Team. *R: A Language and Environment for Statistical Computing*; R Foundation for Statistical Computing: Vienna, Austria, 2019.

42. McMurdie, P.J.; Holmes, S. phyloseq: An R Package for Reproducible Interactive Analysis and Graphics of Microbiome Census Data. *PLoS ONE* **2013**, *8*, e61217. [CrossRef]

43. Kembel, S.W.; Cowan, P.D.; Helmus, M.R.; Cornwell, W.K.; Morlon, H.; Ackerly, D.D.; Blomberg, S.P.; Webb, C.O. Picante: R tools for integrating phylogenies and ecology. *Bioinformatics* **2010**, *26*, 1463–1464. [CrossRef] [PubMed]

44. Battaglia, T. Btools: A Suite of R Function for All Types of Microbial Diversity Analyses. R Package Version 0.0.1. Available online: https://rdrr.io/github/twbattaglia/btools (accessed on 10 July 2019).

45. Kay, M.; Wobbrock, J. ARTool: Aligned Rank Transform for Nonparametric Factorial ANOVAs. R Package Version 0.10.7. Available online: https://github.com/mjskay/ARTool (accessed on 10 July 2019).

46. Lenth, R. Emmeans: Estimated Marginal Means, aka Least-Squares Means. R Package Version 1.4.5. Available online: https://CRAN.R-project.org/package=emmeans (accessed on 10 July 2019).

47. Oksanen, J.; Blanche, F.G.; Friendly, M.; Kindt, R.; Legendre, P.; McGlinn, D.; Minchin, P.R.; O'Hara, R.B.; Simpson, G.L.; Solymos, P.; et al. Vegan: Community Ecology Package. R Package Version 2.5-6. Available online: https://github.com/vegandevs/vegan (accessed on 10 July 2019).

48. Cáceres, M.D.; Legendre, P. Associations between species and groups of sites: Indices and statistical inference. *Ecology* **2009**, *90*, 3566–3574. [CrossRef] [PubMed]

49. Robertson, S.; Ramaley, R.F.; Meyer, T.; Kyndt, J.A. Whole-Genome Sequence of a Unique *Elioraea* Species Strain Isolated from a Yellowstone National Park Hot Spring. *Microbiol. Resour. Announc.* **2019**, *8*, e00907-19. [CrossRef] [PubMed]

50. Yurkov, V.V.; Beatty, J.T. Aerobic Anoxygenic Phototrophic Bacteria. *Microbiol. Mol. Biol. Rev.* **1998**, *62*, 695–724. [CrossRef] [PubMed]

51. Evtushenko, L.; Ariskina, E. Part B. Order XII. *Propionibacteriales* ord. nov., Family II. *Nocardioidaceae*. In *Bergey's Manual*® *of Systematic Bacteriology*; Goodfellow, M., Kämpfer, P., Busse, H.-J., Trujillo, M.E., Suzuki, K., Ludwig, W., Whitman, W.B., Eds.; Springer: New York, NY, USA, 2012; pp. 1189–1197, ISBN 978-0-387-95043-3.

52. Octavia, S.; Lan, R. The Family *Enterobacteriaceae*. In *The Prokaryotes: Gammaproteobacteria*; Rosenberg, E., DeLong, E.F., Lory, S., Stackebrandt, E., Thompson, F., Eds.; Springer: Berlin/Heidelberg, Germany, 2014; pp. 225–286, ISBN 978-3-642-38921-4.

53. *The Prokaryotes: Deltaproteobacteria and Epsilonproteobacteria*; Rosenberg, E.; DeLong, E.F.; Lory, S.; Stackebrandt, E.; Thompson, F. (Eds.) Springer: Berlin/Heidelberg, Germany, 2014; ISBN 978-3-642-39043-2.

54. Marasco, R.; Rolli, E.; Fusi, M.; Michoud, G.; Daffonchio, D. Grapevine rootstocks shape underground bacterial microbiome and networking but not potential functionality. *Microbiome* **2018**, *6*, 3. [CrossRef] [PubMed]

55. Badri, D.V.; Vivanco, J.M. Regulation and Function of Root Exudates. *Plant Cell Environ.* **2009**, *32*, 666–681. [CrossRef] [PubMed]

56. Mönchgesang, S.; Strehmel, N.; Schmidt, S.; Westphal, L.; Taruttis, F.; Müller, E.; Herklotz, S.; Neumann, S.; Scheel, D. Natural variation of root exudates in Arabidopsis thaliana -linking metabolomic and genomic data. *Sci. Rep.* **2016**, *6*, 29033. [CrossRef] [PubMed]

57. Marastoni, L.; Lucini, L.; Miras-Moreno, B.; Trevisan, M.; Sega, D.; Zamboni, A.; Varanini, Z. Changes in Physiological Activities and Root Exudation Profile of Two Grapevine Rootstocks Reveal Common and Specific Strategies for Fe Acquisition. *Sci. Rep.* **2020**, *10*, 18839. [CrossRef]

58. Berlanas, C.; Berbegal, M.; Elena, G.; Laidani, M.; Cibriain, J.F.; Sagües, A.; Gramaje, D. The Fungal and Bacterial Rhizosphere Microbiome Associated with Grapevine Rootstock Genotypes in Mature and Young Vineyards. *Front. Microbiol.* **2019**, *10*, 1142. [CrossRef]
59. Wittebolle, L.; Marzorati, M.; Clement, L.; Balloi, A.; Daffonchio, D.; Heylen, K.; De Vos, P.; Verstraete, W.; Boon, N. Initial community evenness favours functionality under selective stress. *Nature* **2009**, *458*, 623–626. [CrossRef] [PubMed]
60. van der Heijden, M.G.A.; Klironomos, J.N.; Ursic, M.; Moutoglis, P.; Streitwolf-Engel, R.; Boller, T.; Wiemken, A.; Sanders, I.R. Mycorrhizal fungal diversity determines plant biodiversity, ecosystem variability and productivity. *Nature* **1998**, *396*, 69–72. [CrossRef]
61. Wagg, C.; Hautier, Y.; Pellkofer, S.; Banerjee, S.; Schmid, B.; van der Heijden, M.G.A. Diversity and asynchrony in soil microbial communities stabilizes ecosystem functioning. *bioRxiv* **2020**. [CrossRef]
62. Salles, J.F.; Le Roux, X.; Poly, F. Relating Phylogenetic and Functional Diversity among Denitrifiers and Quantifying their Capacity to Predict Community Functioning. *Front. Microbiol.* **2012**, 3. [CrossRef]
63. Jousset, A.; Schulz, W.; Scheu, S.; Eisenhauer, N. Intraspecific genotypic richness and relatedness predict the invasibility of microbial communities. *ISME J.* **2011**, *5*, 1108–1114. [CrossRef]
64. Zarraonaindia, I.; Owens, S.M.; Weisenhorn, P.; West, K.; Hampton-Marcell, J.; Lax, S.; Bokulich, N.A.; Mills, D.A.; Martin, G.; Taghavi, S.; et al. The Soil Microbiome Influences Grapevine-Associated Microbiota. *mBio* **2015**, *6*, e02527-14. [CrossRef] [PubMed]
65. Vink, S.N.; Chrysargyris, A.; Tzortzakis, N.; Salles, J.F. Bacterial Community Dynamics Varies with Soil Management and Irrigation Practices in Grapevines (*Vitis Vinifera* L.). *Appl. Soil Ecol.* **2021**, *158*, 103807. [CrossRef]
66. Baldani, J.I.; Videira, S.S.; dos Santos Teixeira, K.R.; Reis, V.M.; de Oliveira, A.L.M.; Schwab, S.; de Souza, E.M.; Pedraza, R.O.; Baldani, V.L.D.; Hartmann, A. The Family Rhodospirillaceae. In *The Prokaryotes: Alphaproteobacteria and Betaproteobacteria*; Rosenberg, E., DeLong, E.F., Lory, S., Stackebrandt, E., Thompson, F., Eds.; Springer: Berlin/Heidelberg, Germany, 2014; pp. 533–618, ISBN 978-3-642-30197-1.
67. Pujalte, M.J.; Lucena, T.; Ruvira, M.A.; Arahal, D.R.; Macián, M.C. The Family Rhodobacteraceae. In *The Prokaryotes: Alphaproteobacteria and Betaproteobacteria*; Rosenberg, E., DeLong, E.F., Lory, S., Stackebrandt, E., Thompson, F., Eds.; Springer: Berlin/Heidelberg, Germany, 2014; pp. 439–512, ISBN 978-3-642-30197-1.
68. Sakarika, M.; Spanoghe, J.; Sui, Y.; Wambacq, E.; Grunert, O.; Haesaert, G.; Spiller, M.; Vlaeminck, S.E. Purple Non-sulphur Bacteria and Plant Production: Benefits for Fertilization, Stress Resistance and the Environment. *Microb. Biotechnol.* **2020**, *13*, 1336–1365. [CrossRef] [PubMed]
69. Ryan, R.P.; Monchy, S.; Cardinale, M.; Taghavi, S.; Crossman, L.; Avison, M.B.; Berg, G.; van der Lelie, D.; Dow, J.M. The Versatility and Adaptation of Bacteria from the Genus *Stenotrophomonas*. *Nat. Rev. Microbiol.* **2009**, *7*, 514–525. [CrossRef]
70. Liu, D.; Zhang, P.; Chen, D.; Howell, K. From the Vineyard to the Winery: How Microbial Ecology Drives Regional Distinctiveness of Wine. *Front. Microbiol.* **2019**, *10*, 2679. [CrossRef] [PubMed]
71. Bokulich, N.A.; Collins, T.S.; Masarweh, C.; Allen, G.; Heymann, H.; Ebeler, S.E.; Mills, D.A. Associations among Wine Grape Microbiome, Metabolome, and Fermentation Behavior Suggest Microbial Contribution to Regional Wine Characteristics. *mBio* **2016**, *7*, e00631-16. [CrossRef]

Article

Effects of Simulated Nitrogen Deposition on the Bacterial Community of Urban Green Spaces

Lingzi Mo [1,2,3], Augusto Zanella [2], Xiaohua Chen [1], Bin Peng [1], Jiahui Lin [1], Jiaxuan Su [1], Xinghao Luo [1], Guoliang Xu [1,3,*] and Andrea Squartini [4,*]

[1] School of Geography and Remote Sensing, Guangzhou University, WaihuanXi Rd 230, Guangzhou 510006, China; lingzi.mo@phd.unipd.it (L.M.); chenxhua@gzhu.edu.cn (X.C.); 21116010180@e.gzhu.edu.cn (B.P.); 2112001050@e.gzhu.edu.cn (J.L.); 2112001033@e.gzhu.edu.cn (J.S.); 2111801037@e.gzhu.edu.cn (X.L.)

[2] Department of Land, Environment, Agriculture and Forestry TESAF, University of Padova, Viale dell'Università 16, 35020 Legnaro (PD), Italy; augusto.zanella@unipd.it

[3] Rural Non-point Source Pollution Comprehensive Management Technology Center of Guangdong Province, WaihuanXi Rd 230, Guangzhou 510006, China

[4] Department of Agronomy, Animals, Food, Natural Resources and Environment DAFNAE, University of Padova, Viale dell'Università 16, 35020 Legnaro (PD), Italy

[*] Correspondence: xugl@gzhu.edu.cn (G.X.); squart@unipd.it (A.S.)

Abstract: Continuing nitrogen (N) deposition has a wide-ranging impact on terrestrial ecosystems. To test the hypothesis that, under N deposition, bacterial communities could suffer a negative impact, and in a relatively short timeframe, an experiment was carried out for a year in an urban area featuring a cover of Bermuda grass (*Cynodon dactylon*) and simulating environmental N deposition. NH_4NO_3 was added as external N source, with four dosages (N0 = 0 kg N ha^{-2} y^{-1}, N1 = 50 kg N ha^{-2} y^{-1}, N2 = 100 kg N ha^{-2} y^{-1}, N3 = 150 kg N ha^{-2} y^{-1}). We analyzed the bacterial community composition after soil DNA extraction through the pyrosequencing of the 16S rRNA gene amplicons. N deposition resulted in soil bacterial community changes at a clear dosage-dependent rate. Soil bacterial diversity and evenness showed a clear trend of time-dependent decline under repeated N application. Ammonium nitrogen enrichment, either directly or in relation to pH decrease, resulted in the main environmental factor related to the shift of taxa proportions within the urban green space soil bacterial community and qualified as a putative important driver of bacterial diversity abatement. Such an impact on soil life induced by N deposition may pose a serious threat to urban soil ecosystem stability and surrounding areas.

Keywords: Nitrogen deposition; bacteria; soil biodiversity; urban; 16S rRNA

Citation: Mo, L.; Zanella, A.; Chen, X.; Peng, B.; Lin, J.; Su, J.; Luo, X.; Xu, G.; Squartini, A. Effects of Simulated Nitrogen Deposition on the Bacterial Community of Urban Green Spaces. *Appl. Sci.* **2021**, *11*, 918. https://doi.org/10.3390/app11030918

Received: 9 December 2020
Accepted: 18 January 2021
Published: 20 January 2021

Publisher's Note: MDPI stays neutral with regard to jurisdictional claims in published maps and institutional affiliations.

1. Introduction

Increasing nitrogen (N) deposition caused by industrialization and anthropogenic activities has become an ecological problem that attracted worldwide attention [1–3]. The terrestrial ecosystems in recent decades have suffered increases in anthropogenic inputs of pollutants and increases in the deposition of ammonia (NH_3), nitrogen oxides (NO_x), and their reactive products (NH_4^+-N, NO_3^--N, and HNO_3) [4]. Especially in China, atmospheric N deposition has increased by 8 kg N ha^{-1} since the 1980s [5]. Sustained N deposition has a wide-ranging impact on terrestrial ecosystems [6], which can cause soil acidification, reduction in biological functions, and diversity [4,7].

With the fast-growing urbanization, city areas are the most intense zones of human activity. Urbanization can alter the abiotic and biotic soil environment through several means, including atmospheric deposition, urban heat island effect, and invasion of exotic species, which may have contrasting effects on microbial activity and function [8]. High population density, heavy traffic, industrial, and agricultural impacts can cause high N emissions in urban areas [9]. The sources of N falling out on urban are complex.

Some studies found that nitrogen oxides produced in cities can return to urban areas through atmospheric deposition [10], which ultimately became a source of nutrients for soil biota in urban ecosystems [11]. Atmospheric N deposition can change the structure, function, and processes of ecosystems [12,13]. Therefore, sustained N deposition may affect the complexity of the entire ecological environment, including the urban green spaces, which are particularly prone to receive an impact, since those are the areas that allow concentrated water infiltration, being the sole permeable points among a system of paved surfaces and buildings, and being located usually in contexts suffering massive pollution fallout. As an essential part of the city environment, urban green space provides a variety of ecosystem services [14], such as contaminant degradation [15], carbon and nitrogen nutrient cycling [16], and general biochemical cycling [17]. It constitutes an interactive arterial system that helps to improve the quality of life of urban residents [18]. A better understanding of urban green space functionality, in response to external inputs in the urban soil ecosystem, is fundamental to improve urban soil management, and, in particular, nitrogen requirements, and to reduce threats to the environment.

The functioning of a soil depends on a sophisticated, but the coordinated system of interactions among the environment, vegetation, and organisms living in the soil [19–23]. Large in quantity, and extreme in diversity, soil microbes are important participants in the soil ecosystem and play a key role in biogeochemical cycling [24]. They assist the ecological processes of litter decomposition, soil fertility formation and maintenance, and nutrient cycling [25,26]. Soil microorganisms are commonly used to indicate the environmental changes of soil because they are sensitive to the changes in environmental factors (e.g., soil nutrients and pH) [27]. In recent years, with the development of high-throughput sequencing techniques, research methods for the analysis of environmental microorganisms have entered their most advanced stage [28]. Metagenomics is the study of a living community by direct extraction and subsequent reading of DNA sequences from all of the different genomes of organisms hosted in a given space. It can be used to understand the complete structure of microbial communities, revealing their environmental diversity and complexity [29,30]. Studies of the microbial diversity of soil, water, wetlands, and extreme environmental by means of DNA sequences have increased lately [31–34].

Other analyses have confirmed that there are a large number of bacteria that find their critical habitats in soil, and that their activities are strictly related to underground processes [35]. Bacteria that adjust to living in the soil can also affect plant growth and health through their direct and indirect effects in ecological processes in relation to N cycling [36]. Studying the feedback of urban soil bacteria to N deposition is therefore of primary importance to understand the effects of regional environmental changes on ecosystem processes regulated by soil bacteria. Although the structure of bacterial communities entails particular importance for the stability and productivity of urban green spaces, there are still few studies on the effect of anthropogenic N deposition on these habitats.

In this respect, since (a) nitrogen is the primary nutrient for soil fertilization and (b) life is N-dependent and N-limited in almost all environments, one could not predict a priori whether N deposition would be deleterious for microbial ecology, or it could instead boost its development and diversification, as is the case for crops and weeds. The hypothesis we aimed at testing here was that, in spite of its nature of essential macro-element for nutrition, sustained N additions to a soil could lead to a fast depression of the main ecological indicators for bacterial diversity in soil. To unveil the potential effects of different degrees of N deposition on the bacterial communities of urban green areas, we set up a simulated atmospheric N deposition-controlled experiment. Lawns, planted with *Cynodon dactylon*, which is one of the most popularly used grasses for turf, yards, and recreational greens, were selected as the research object. We analyzed the bacterial diversity and community structure of urban green space undergoing deliberate supplementation of N deposition by Illumina MiSeq 16S NGS metabarcoding. The present study aimed at (1) detecting the dynamics of the urban green space bacterial diversity along an N loading gradient;

and (2) investigating the direct and indirect effects of N deposition on soil microbial communities in urban green spaces.

2. Materials and Methods

2.1. Study Sites and Experimental Design

The study was carried out in a green space experimental field at Guangzhou University in Guangzhou, China ($113°23'$ E, $23°02'$ N). The climate of the area is a typical subtropical monsoon climate, with warm and wet summer, and short and dry winter. The annual average temperature is 21.5 °C, and the average relative humidity is 77%. The hottest month is July, with an average monthly temperature of 28.7 °C. The coldest month is January, and the monthly average temperature is 13.3 °C. The rainy season is from April to September. The N deposition inputs in the four Guangzhou districts (Baiyun, Tianhe, Luogang, and Conghua) average 43.3, 41.2, 35.2, and 30.1 kg N ha^{-1} y^{-1}, respectively [37].

The experiment started in May 2016 and continued until May 2017. A randomized block design was used, with four treatments and five replicate plots of each treatment. Twenty 2 m $\times$ 2 m plots were arranged in a 4 $\times$ 5 matrix; 50 g of *Cynodon dactylon* seeds were evenly planted in each plot. The distance between any two adjacent plots was 0.5 m. The chemical parameters of the chosen soil at time zero, before the beginning of the trial (means of four replicates $\pm$ standard deviation) were the following; pH: 8.28 $\pm$ 0.06; ammonium nitrogen (mg kg^{-1}): 1.54 $\pm$ 0.10; nitrate nitrogen (mg kg^{-1}): 3.37 $\pm$ 0.03; total nitrogen (g kg^{-1}): 0.61 $\pm$ 0.06; total carbon (g kg^{-1}): 13.41 $\pm$ 0.96. NH$_4$NO$_3$ was added as the external N sources once a month, and the final N input corresponded to N0 (0 kg N ha^{-2} y^{-1}), N1 (50 kg N ha^{-2} y^{-1}), N2 (100 kg N ha^{-2} y^{-1}), N3 (150 kg N ha^{-2} y^{-1}). The experimental field was fenced off to prevent disturbance.

2.2. Soil Sampling and Analysis

Three months (July 2016), 6 months (October 2016), and 12 months (May 2017) after the start of the N supplementation treatment, three subsamples were sampled at a 0–5 cm depth of every plot. Subsequently, subsamples were mixed thoroughly and then separated into two portions. One portion of the soil was immediately frozen in liquid nitrogen and then stored at −80 °C for genomic DNA extraction. The remaining portion of soil was passed through a 2 mm sieve and used for soil chemical analysis. A total of 20 samples represented the five replicates of four treatments at every sampling time.

Soil pH (water: soil, 2.5:1) was determined using a pH meter. Nitrate (NO$_3$$^-$-N) and ammonium (NH$_4$$^+$-N,) were extracted from 10 g of soil using 2 M KCl and determined using a San^{++} Continuous Flow Analyzer (Skalar, Breda, The Netherlands). Soil total carbon (TC) was determined by the dichromate digestion method, and total nitrogen (TN) was measured by the Kjeldahl method [38].

2.3. DNA Extraction

Total metagenomic DNA was extracted and purified from samples with a Universal Genomic DNA Kit (CwBio Inc., Beijing, China) according to the manufacturer's instructions. The concentration of total DNA was measured using the Qubit Platform (Life Technologies, CA, USA). The V3–V4 hypervariable region of 16S ribosomal RNA gene was amplified from 30 ng of DNA in triplicate polymerase chain reactions (PCR) with Pyro best DNA polymerase (TaKaRa, Dalian, China) and the following forward and reverse primers, respectively, 5′-ACTCCTACGGGAGGCAGCAG-3′ and 5′-GGACTACHVGGGTWTCTAAT-3′ (Sangon Biotech, Shanghai, China), following the manufacturer's instructions. Barcode sequences were attached to the amplification primers to distinguish the different samples. PCR products were inspected on 2% agarose electrophoresis gel and the respective amplicon libraries generated. Microbial DNA was then sequenced by Honortech (Beijing, China) using the Illumina MiSeq platform. Based on the raw data, pair-end reads were spliced using the principle of 98% overlap of 19 bases using the Connecting Overlapped Pair-End software 36. Barcode and primer sequences were then filtered to obtain the clean data. Operational Taxonomic Units (OTUs) were individuated as those clustering at shared nucleotide identity equal or higher than 97%. The relative abundance of soil bacteria was calculated according to the species annotation and reads number.

2.4. Statistical Analysis

Alpha diversity was determined based on the Chao1, and by Shannon–Wiener indices. Chao1 Index is based on the number of OTUs with an individual sequence called "singletons" and the number of OTUs containing a pair of sequences is called "doubletons". The Chao1 index being more sensitive to rare species in the community [39]. The Chao1 were calculated by using the following equations:

$$\text{Chao1} = S_{OTU} + \frac{F_1^2}{2F_2} \tag{1}$$

where S_{otu}, F_1 and F_2 represent the number of observed species, singletons, and doubletons

Shannon index uses the number of sequences in each OTU and the total number of sequences in the community for calculation. The Shannon Index estimate is given by:

$$\text{Shannon index} = -\sum_{i=1}^{s} p_i \ln p_i \tag{2}$$

where P_i is the proportion (n/N) of individuals of one particular species found (n) divided by the total number of individuals found.

Soil properties, bacterial abundance, and alpha diversity index data were analyzed by one-way analysis of variance (ANOVA) to determine significant differences among the treatments. Associations between bacterial community composition and a single environmental variable (N deposition) were identified by preliminary verification of variance homogeneity and application of the Anderson's PERMDISP2 procedure, (http://scikit-bio. org/docs/0.5.4/generated/generated/skbio.stats.distance.permdisp.html), which visualizes the distances of each sample to the group centroid in a PCoA and provides a *p*-value for the significance of the grouping

The Pearson correlation coefficient was used to test the relationships between soil properties and alpha diversity indexes. The Spearman's rank correlation coefficient was utilized to assess the correlations between soil properties and bacterial main phyla, as in those cases, some of the data did not comply with distribution normality, calling for a non-parametric method. The correlation between soil properties and bacterial community structure was further investigated by the Mantel test. Pearson and Spearman's correlation tests were performed using SPSS 18.0 (SPSS Inc., Chicago, IL, USA). ANOVA, PERMDISP2, and Mantel tests were performed using R software (Version 3.6.1). PERMDISP2 is implemented in R using the vegan betadisper function (version 2.4-2, License: GPL-2).

3. Results

3.1. Effects of N Deposition on Soil Properties

Inorganic nitrogen dosage markedly affected most of the measured soil properties, such as pH, NH_4^+-N, NO_3^--N, total nitrogen (TN), and total carbon (TC) contents (Table 1). After 12 months of N deposition, soil pH under N3 treatment was significantly lower than that of the N0 ($p < 0.01$) and N1 ($p < 0.05$) plots. Moreover, the pH under the N2 treatment was significantly lower than that of the N0 control plot ($p < 0.05$). The contents of NH_4^+-N and NO_3^--N under the N2 and N3 treatments were significantly higher than those of the N0 and N1 treatments ($p < 0.01$). While total nitrogen (TN) and total carbon (TC) did not show significant differences ($p > 0.05$) within the N gradient of the same time point, changes were nevertheless occurring in a time-dependent fashion with a steady decrease of TC and an increase of TN.

Table 1. Effects of N deposition on soil properties.

Time	Treatment	pH	NH$_4^+$-N (mg kg^{-1})	NO$_3^-$-N (mg kg^{-1})	TN (g kg^{-1})	TC (g kg^{-1})
3 months	N0	7.89 ± 0.18 [a]	2.95 ± 0.54 [a]	27.712 ± 18.84 [a]	0.78 ± 0.09 [a]	12.77 ± 1.39 [a]
	N1	7.77 ± 0.29 [a]	3.50 ± 0.40 [a]	11.05 ± 5.37 [a]	0.82 ± 0.14 [a]	13.96 ± 1.23 [a]
	N2	7.58 ± 0.29 [a]	2.23 ± 0.57 [a]	42.04 ± 15.71 [a]	0.72 ± 0.07 [a]	12.93 ± 1.14 [a]
	N3	7.50 ± 0.43 [a]	3.03 ± 0.24 [a]	19.68 ± 5.79 [a]	0.67 ± 0.05 [a]	12.14 ± 1.25a
	p *	0.81	0.31	0.40	0.66	0.78
6 months	N0	7.92 ± 0.13 [a]	7.49 ± 0.12 [b]	0.59 ± 0.03 [d]	0.64 ± 0.12 [a]	10.88 ± 0.45 [a]
	N1	7.22 ± 0.45 [a,b]	15.40 ± 6.28 [b]	4.22 ± 0.32 [c]	0.78 ± 0.18 [a]	12.61 ± 1.66 [a]
	N2	6.56 ± 0.23 [b]	31.61 ± 3.80 [b]	9.29 ± 1.04 [b]	0.82 ± 0.10 [a]	11.34 ± 0.84 [a]
	N3	6.23 ± 0.47 [b]	75.26 ± 17.67 [a]	15.29 ± 1.81 [a]	0.94 ± 0.09 [a]	10.76 ± 0.41 [a]
	p *	<0.05	<0.01	<0.01	0.45	0.54
12 months	N0	7.79 ± 0.08 [a]	19.34 ± 0.74 [b]	1.48 ± 0.38 [b]	0.82 ± 0.08 [b]	8.78 ± 0.81 [a]
	N1	7.08 ± 0.47 [a,b]	37.34 ± 9.41 [b]	6.49 ± 0.99 [b]	0.88 ± 0.07 [b]	8.90 ± 0.74 [a]
	N2	6.23 ± 0.47 [b,c]	78.94 ± 11.71 [a]	17.13 ± 2.45 [a]	1.12 ± 0.23 [a,b]	8.66 ± 1.31 [a]
	N3	5.73 ± 0.47 [c]	108.69 ± 18.90 [a]	23.18 ± 4.13 [a]	1.33 ± 0.12 [a]	8.60 ± 0.34 [a]
	p *	<0.05	<0.01	<0.01	0.08	0.99

Data are presented as mean and standard errors (in brackets of the five replicates; different letters indicate significant differences between treatments *: *p*-values are based on ANOVA.

3.2. Effects of N Deposition on Soil Bacterial Richness and Diversity

With the increase of N deposition in time and dosages, ecological parameters, such as richness and diversity of soil bacteria, decreased (Figures 1 and 2). At 6 months, soil bacterial richness and diversity were lower than those at 3 months, but the difference among the four dosages was not significant. After 12 months of N treatment, richness and diversity showed an ordered array N3 < N2 < N1 < N0, and there was a significant difference between N3 and N2 treatments ($p < 0.05$), and a highly significant difference between N0 and N1 treatments ($p < 0.01$).

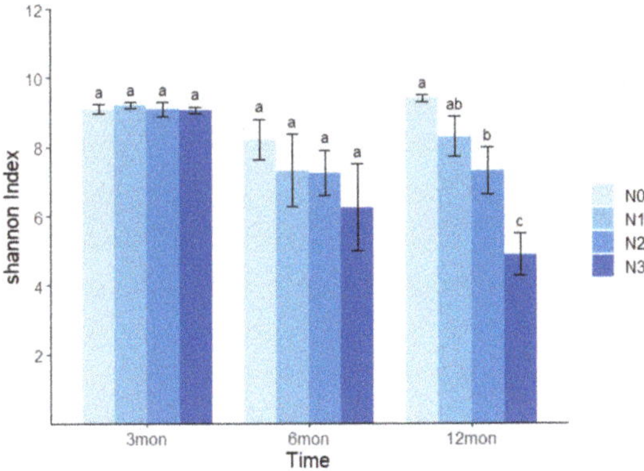

Figure 1. Effects of N deposition on soil bacterial Shannon diversity index. Different letters indicate significant differences among treatments within the same sampling time ($p < 0.05$, ANOVA).

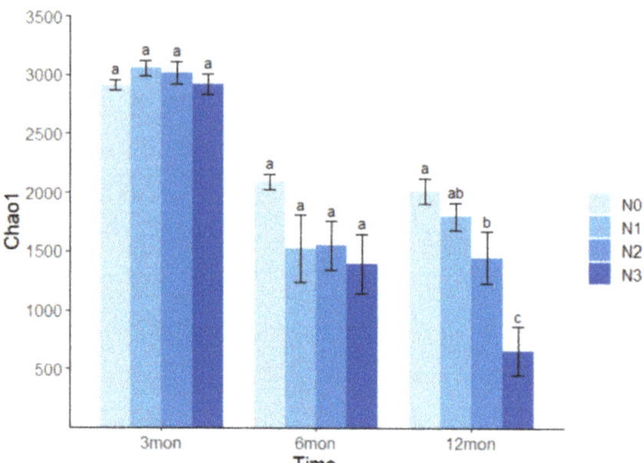

Figure 2. Effects of N deposition on soil bacterial chao1 index. Different letters indicate significant differences among treatments within the same sampling time ($p < 0.05$, ANOVA).

3.3. Effects of N Deposition on Soil Bacterial Community Composition

Upon reads processing ad filtering the actual sequences that were obtained in the bacterial community analysis of three sampling times were 530,447; 960,673; and 522,682; respectively. Classified OTUs belonged to 51 phyla among all samples. As common, due to our still partial knowledge of global microbial diversity, available reference databases do not allow an unambiguous assignment up to the genus or species level for a substantial number of the obtained sequence reads. Therefore, the most reliable annotation is to date prudentially assumed at relatively high taxonomy ranks. For this reason, we will present to the majority of results sticking to the phylum level, and we will later comment the data at finer resolution (from order to genus level) for those cases in which the assignment was backed up by a robust degree of sequence identity.

The bacterial community composition at phylum level (relative abundance >1%) is shown in Figure 3. The dominant phyla (i.e., the ones showing highest values in relative sequence abundance) in all samples were Proteobacteria and Acidobacteria. Besides, the subdominant phyla in all samples were Chloroflexi, Bacteroidetes, Nitrospirae, Acidobacteria, Gemmatimonadetes, Verrucomicrobia, Planctomycetes, Firmicutes, Actinobacteria, Parcubacteria, and Latescibacteria. These phyla represented more than 88% of the sequences in all the samples. After 12 months of N treatment, the relative abundance of Proteobacteria significantly differed among different treatments ($p < 0.01$, Figure S1 in the Supplementary Material), specifically increasing following N enrichment. The percentage of Acidobacteria was low in the N3 treatment but did not differ significantly among the treatments (Figure S1).

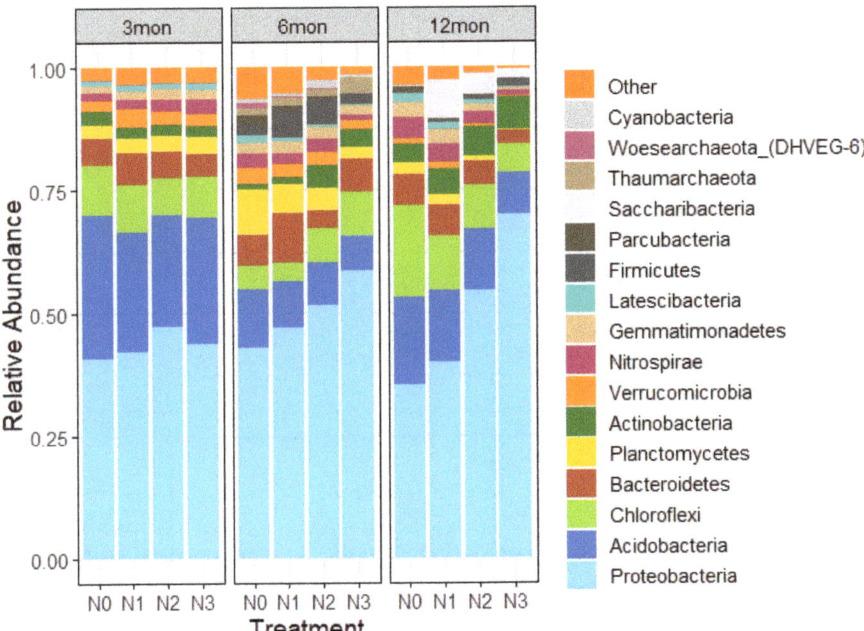

Figure 3. Effects of N deposition on soil bacterial relative abundance. The groups accounting for ≥1% are shown while those >1% are integrated into "other".

The variations in soil bacterial communities caused by N deposition were evaluated by PERMDISP2. Differences in the overall bacterial composition were expressed based on the Bray–Curtis distance. Results are shown in Figure 4. After 3 months (July 2016) of N deposition, different N treatments did not separate, which indicated that at such early time, the bacterial community composition was still similar among all treatments (Figure 4). Subsequently, starting from the second time point (October 2016), a significant difference arose between communities in relation to the N level received. After 12 months (May 2017) of N deposition, the communities separate clearly between N3 and N0 treatments (Figure 4), and the bacterial community composition of N3 treatment resulted profoundly changed. Such significance attains its highest value in the last sampling at 12 months (May 2017). An N deposition rate of N3 or higher was required for a significant shift to occur.

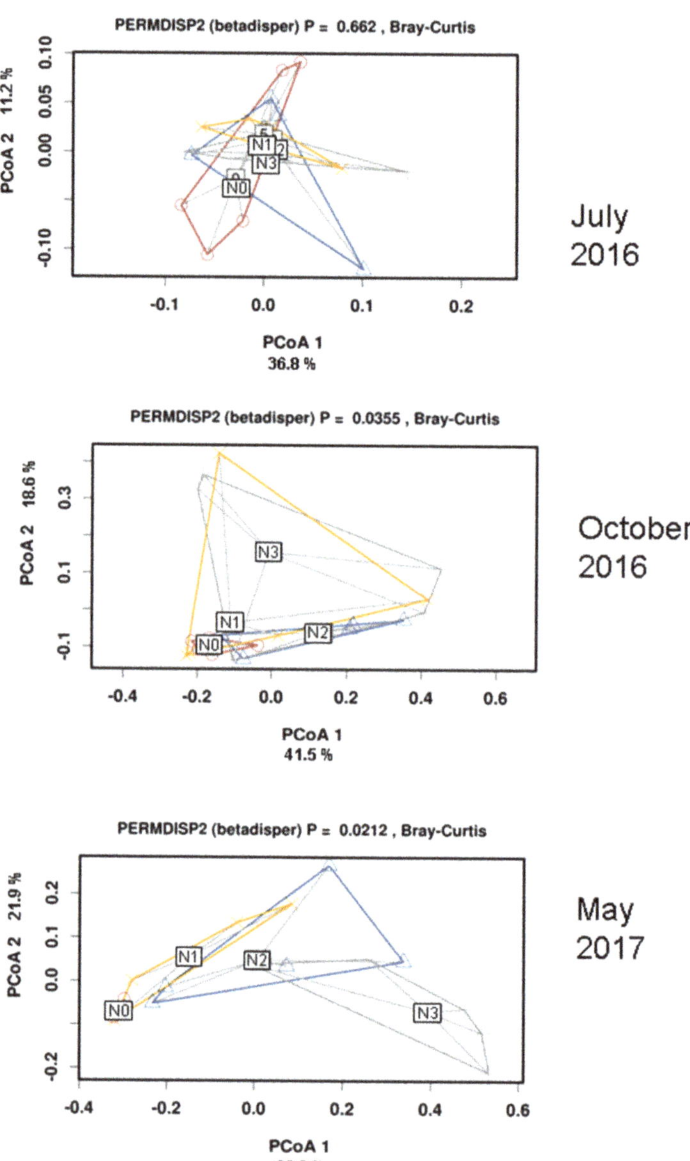

Figure 4. Differences in bacterial community composition among N treatments across different sampling times. PERMDISP2 visualizes the distances of each sample to the group centroid in a principal coordinate analysis (PCoA) and provides a *p*-value for the significance of the treatments.

3.4. Effects of Soil Chemical Properties on the Soil Bacterial Diversity and Community Composition

The correlations of bacterial diversity indices and main phyla with soil properties are shown in Figure 5. Values of the Chao1 index were positively correlated with soil pH ($p < 0.001$) and TC contents ($p < 0.001$) and were significantly negatively correlated with NH_4^+-N ($p < 0.001$), and TN contents ($p < 0.05$). Values of the Shannon index were positively correlated with soil pH ($p < 0.001$) and TC contents ($p < 0.05$) and were significantly negatively correlated with NH_4^+-N ($p < 0.001$). The majority of the predominant phyla showed a significant correlation with certain soil chemical properties, whereas only the phyla. The relative abundances of Proteobacteria were negatively correlated with soil pH. Actinobacteria and Firmicutes were also negatively correlated with soil pH, whereas they showed a positive correlation with the NH_4^+. The abundances of Acidobacteria, Planctomycetes, Latescibacteria, Verrucomicrobia, and Parcubacteria presented a positive relationship with pH, while they showed a negative relationship with NH_4^+-N. Nitrospirae and Gemmatimonadetes did not exhibit a significant correlation with any of the soil parameters.

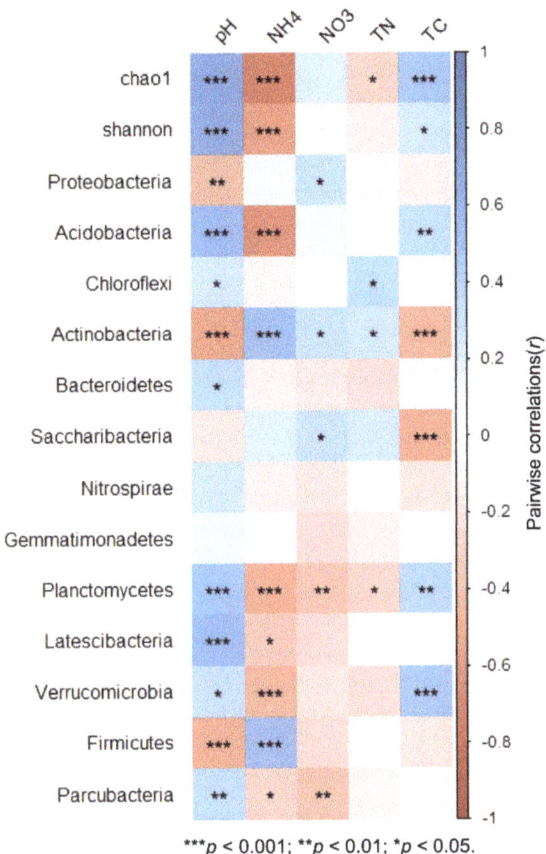

$***p < 0.001$; $**p < 0.01$; $*p < 0.05$.

Figure 5. Pairwise correlations of bacterial diversity indices and abundant phyla (relative abundance >1%) with soil properties. The Pearson correlation coefficient was used to test the relationships between soil properties and alpha diversity indexes. Spearman correlation coefficients were utilized to assess the correlations between soil properties and bacterial main phyla.

The Mantel test explored the correlation between bacterial communities and soil properties (Table 2). The correlation coefficients, in decreasing order of absolute values, were, for positive correlations: pH, NH_4^+-N, and TN; while for negative correlations: TC, and NO_3^--N. Two of the most important factors, pH ($p < 0.05$) and NH_4^+-N ($p < 0.05$), were closely correlated with bacterial community composition. The non-opposite trend for pH and ammonium appearing from Table 2 is, in our view, due to the fact that, although ecological diversity indexes are abated by increasing ammonium, and decreasing pH, a number of numerically relevant phyla are enhanced. Thus, the overall value of community structure is not as impaired by such substitutions in relative abundance as the Simpson and evenness parameters are.

Table 2. Correlations among the overall bacterial community structure and soil proprieties according to the Mantel test.

Soil Properties	r	p
pH	0.2273	0.004
NH_4^+-N	0.1799	0.037
NO_3^--N	−0.1457	0.986
TN	0.0066	0.401
TC	−0.0815	0.848

It must be observed however that in Table 2, while for the pH parameter, the p value is very significant, for the ammonium nitrogen, significance is borderline, and the r value is very low.

4. Discussion

4.1. Effects of Soil Chemical Properties on the Soil Bacterial Diversity and Community Composition

To clarify the global interactions between urban soils and their inhabiting microbes, we deemed necessary to understand the specific ones occurring between urban green space soil and hosted bacteria, in order to seize the effects of N enrichment on the whole urban ecosystem functioning. Sequencing analyses revealed that Proteobacteria and Acidobacteria were the dominant phyla in all samples. When considering the time and dose-related patterns, Proteobacteria specifically increased following N enrichment, while Acidobacteria, Chloroflexi, Bacteroidetes, and Nitrospirae exhibited the opposite trend. Therefore, the response of the bacterial community to N deposition appears to vary substantially among different phyla.

The deposition of N as a nutrient may trigger both direct and indirect effects on microbial communities and diversity [40,41]. These can affect the soil nutrient pool, along with environmental factors changes. N deposition could affect soil bacteria by changing soil inorganic N content. In addition to soil N content, pH is one of the critical factors affecting soil microbial diversity [42,43], because N could indirectly affect soil bacterial community by soil acidification [44,45]. It is, in fact, widely accepted that soil acidification can directly influence soil bacterial community composition [43,46]. In this sense, most of the prior studies indicated that soil pH was more important than nutrients in shaping bacterial community structure [47–50].

In our experiments, we observed that the response of urban green space bacteria to N enrichment appears related mainly to NH_4^+-N and to its effects on soil pH (Figure 5). The soil dissolved inorganic N (NH_4^+-N and NO_3^--N were generally correlated with the N treatment gradient, which was accompanied by a corresponding decrease in soil pH. The reason for pH decreases as a consequence of ammonium-based fertilization is mainly to be seen in the fact that when ammonium undergoes nitrification it leaves net H^+ ions into the circulating solution. Bacterial sensitivity to acidic conditions is widespread and mainly due to the scarce attitude to maintain neutrality within the cytoplasm, to avoid amino acid charged groups depolarization.

From the statistical point of view, significant differences resulted in soil pH, NH_4^+-N and NO_3^--N among different N treatments after 12 months of experimentation (Table 1). Considering the different forms of N, both NH_4^+-N and NO_3^--N can cause soil acidification [51]. Our data showed that soil bacterial community composition was closely related to the soil chemical parameters. In fact, some of the parameters did correlate and they overlapped in their explanatory power under N enrichment. The correlations of phylum-level abundances with soil pH and NH_4^+-N were observed for most of the dominant phyla (Figure 5). In particular, our results suggest that the bacterial community structure has a significant relationship with soil pH and NH_4^+-N (Table 2). Soil mineral N availability was supposedly also tightly connected to the bacterial community structure as other studies reported [50,52]. Besides, Nie et al. pointed out that ammonium nitrogen content was a dominant predictor of bacterial community composition in acidic soil with exogenous nitrogen enrichment [53]. An alternative hypothesis underlying the response of dominant phyla to the N addition is the functional classification model. Proteobacteria and Actinobacteria, that have fast growth rates, were more likely to increase in nutrient-rich conditions, while Acidobacteria and Chloroflexi that have slower growth rates, would likely decline [35,54]. In general, Acidobacteria are adapted to low pH conditions [55]. However, after 12 months of added N, Proteobacteria were enriched under high N treatment, while interestingly, Acidobacteria declined. N2 and N3 treatments were shown to lower pH but, in comparison to the N0 treatment, Acidobacteria did not increase (Figure 3). The reasons for this behavior could be sought in the fact that effects of pH on the different groups of Acidobacteria can be very different [56]. In this sense, as confirmed by our study, pH might not necessarily be the main driver of changes for the Acidobacteria phyla as a whole. Moreover, correlations between soil pH and bacterial communities following N deposition do not necessarily demonstrate nor imply their direct causative relation. In comparison, soil ammonium nitrogen was an important environmental factor that appeared to explain many of the changes in the soil bacterial community structure observed in the present study. Specifically, N deposition directly affected urban green space soil bacterial communities in relation to the increase in soil ammonium.

However, we noticed that also the ammonium in the N0 plot increased along time. Atmospheric N deposition is closely related to precipitation [57]. Previous studies have shown Guangzhou's inorganic N (NH_4^+-N and NO_3^--N) coming as precipitation displays a seasonal fluctuation, and NH_4^+-N was the primary form of N deposition in Guangzhou [58]. In 2011, the NH_4^+-N content in precipitation was 13.876 kg ha^{-1}, and NO_3^--N was 6.562 kg ha^{-1}. Especially in summer, NH_4^+-N in precipitation reach 951.44 mg.m^{-2} 9.514 kg ha^{-1}, while NO_3^--N was 4.484 kg ha^{-1} [59]. Since our experiments were set in open air, and were therefore subjected to natural N deposition, this background can explain the observed rise of N also in the control plot. As regards the drop of total carbon that occurs in time, this can be seen as a consequence of the increase of N which is therefore no longer a limiting factor for carbon consumption and assimilation by soil biota, thus lowering the C/N ratio and pushing more carbon towards its mineralization and volatilization as CO_2.

4.2. Temporal Trends and Potential Consequences of N Deposition

The present study focused on the delicate areas of the urban greens in the Chinese atmospheric fallout is consistent with previous ones that showed the decline of soil bacterial biodiversity under N enrichment [40,59–61]. Considering the three observation periods separately, the short-term (3 months) N deposition did not change the composition of soil bacterial diversity and communities (Figures 1, 2 and 4). The response of soil bacterial alpha diversity and community composition to N deposition was gradually confirmed after N addition for 6 months and 12 months (Figures 1, 2 and 4). Results indicate that high deposition (100 and 150 kg N ha^{-2} y^{-1}) caused the critical changes in soil bacterial of urban green space. Experiments that use a high rate of N deposition may be useful for assessing the long-term effects of chronic low rates of N deposition [62]. The results of this study indicate that the higher deposition or accumulation of N in the urban green space deeply affects patterns of bacterial diversity and community structure. The bacterial response is affected by the amount of added N as well as by the duration of the treatment. Overall, nitrogen (N) addition had a negative impact on bacterial richness and diversity in the urban green space in both dose- and time-dependent fashions.

At finer taxonomy level, it is worth reporting that one of the genera that markedly increased, as N was added, was *Mizugakiibacter*, a genus known to be a denitrified, which is expected as a consequence of the added nitrate and the oxidation of ammonium. On the opposite side, the main decline regarded the *Nitrosomonadales* order, which is consistent with the fact that nitrifiers are made superfluous by our addition of N forms, which are already oxidized, as nitrate is. We must not underestimate that N deposition causes changes in the soil environment because it does change bacterial composition at finer taxonomic levels other than phylum, and most of all that it heavily impacts on overall taxa diversity and richness. If N deposition keeps increasing in the future, the decline of bacterial diversity may become dramatically worse. Soil microbial diversity plays a central role in ecosystem processes by driving the Earth surface's biogeochemical cycles [63,64], and therefore any changes in soil bacteria by N deposition might affect not only urban ecosystem processes and functions, but eventually the global ones.

5. Conclusions

High N deposition rate resulted in significant changes in bacterial community composition and showed a major loss of bacterial diversity in soil. The different responses of major phyla to N enrichment appear to be the main reason for the change in the overall bacterial community composition. pH and ammonium resulted, themselves, correlated and overlapped in their explanatory power under N deposition. Apparently, N deposition directly affected bacterial community composition by increasing soil ammonium content. The changes in bacterial community composition and diversity were tightly connected with the amounts of added N, as well as with the duration of the treatment. The decrease in biodiversity induced by N deposition may pose a serious threat to urban soil ecosystem stability, which emphasizes the necessity of thorough and concerted studies to prompt adequate policies to counteract these globally increasing threats.

Supplementary Materials: The following are available online at https://www.mdpi.com/2076-3417/11/3/918/s1, Figure S1: Effects of N deposition on the relative abundance of dominant phyla after 12 months. Different letters indicate significant differences among treatments within the same sampling time ($p < 0.05$, ANOVA).

Author Contributions: Conceptualization, L.M., G.X., X.C. and B.P.; formal analysis, L.M., A.S., B.P., J.L. and J.S.; methodology, L.M., G.X., X.C., B.P., J.L., J.S. and X.L.; writing—original draft, L.M.; writing—review and editing, L.M., A.S., A.Z. and G.X.; funding acquisition, G.X. All authors have read and agreed to the published version of the manuscript.

Funding: This research was funded by National Natural Science Foundation of China, grant number 42071061, 41571247 and U1701236.

Institutional Review Board Statement: Not applicable.

Informed Consent Statement: Not applicable.

Data Availability Statement: The data presented in this study are available on request from the corresponding author.

Acknowledgments: We thank Guangzhou University for the scholarship.

Conflicts of Interest: The authors declare no conflict of interest.

References

1. Liu, X.; Duan, L.; Mo, J.; Du, E.; Shen, J.; Lu, X.; Zhang, Y.; Zhou, X.; He, C.; Zhang, F. Nitrogen deposition and its ecological impact in China: An overview. *Environ. Pollut.* **2011**, *159*, 2251–2264. [CrossRef]
2. Bobbink, R.; Hicks, K.; Galloway, J.; Spranger, T.; Alkemade, R.; Ashmore, M.; Bustamante, M.; Cinderby, S.; Davidson, E.; Dentener, F.; et al. Global assessment of nitrogen deposition effects on terrestrial plant diversity: A synthesis. *Ecol. Appl.* **2010**, *20*, 30–59. [CrossRef] [PubMed]
3. Boutin, M.; Corcket, E.; Alard, D.; Villar, L.; Jiménez, J.J.; Blaix, C.; Lemaire, C.; Corriol, G.; Lamaze, T.; Pornon, A. Nitrogen deposition and climate change have increased vascular plant species richness and altered the composition of grazed subalpine grasslands. *J. Ecol.* **2017**, *105*, 1199–1209. [CrossRef]
4. Galloway, J.N.; Dentener, F.J.; Capone, D.G.; Boyer, E.W.; Howarth, R.W.; Seitzinger, S.P.; Asner, G.P.; Cleveland, C.C.; Green, P.A.; Holland, E.A. Nitrogen cycles: Past, present, and future. *Biogeochemistry* **2004**, *70*, 153–226. [CrossRef]
5. Liu, X.; Zhang, Y.; Han, W.; Tang, A.; Shen, J.; Cui, Z.; Vitousek, P.; Erisman, J.W.; Goulding, K.; Christie, P. Enhanced nitrogen deposition over China. *Nature* **2013**, *494*, 459. [CrossRef]

6. Li-Hua, T.; Hong-Ling, H.; Ting-Xing, H.; Zhang, J.; Li, L.; Ren-Hong, L.; Hong-Zhong, D.; Shou-Hua, L. Decomposition of different litter fractions in a subtropical bamboo ecosystem as affected by experimental nitrogen deposition. *Pedosphere* **2011**, *21*, 685–695.

7. Vitousek, P.M.; Aber, J.D.; Howarth, R.W.; Likens, G.E.; Matson, P.A.; Schindler, D.W.; Schlesinger, W.H.; Tilman, D.G. Human alteration of the global nitrogen cycle: Sources and consequences. *Ecol. Appl.* **1997**, *7*, 737–750. [CrossRef]

8. Enloe, H.A.; Lockaby, B.G.; Zipperer, W.C.; Somers, G.L. Urbanization effects on soil nitrogen transformations and microbial biomass in the subtropics. *Urban Ecosyst.* **2015**, *18*, 963–976. [CrossRef]

9. Templer, P.H.; Weathers, K.C.; Lindsey, A.; Lenoir, K.; Scott, L. Atmospheric inputs and nitrogen saturation status in and adjacent to Class I wilderness areas of the northeastern US. *Oecologia* **2015**, *177*, 5–15. [CrossRef]

10. Russell, A.G.; Winner, D.A.; Harley, R.A.; McCue, K.F.; Cass, G.R. Mathematical modeling and control of the dry deposition flux of nitrogen-containing air pollutants. *Environ. Sci. Technol.* **1993**, *27*, 2772–2782. [CrossRef]

11. Schaefer, S.C.; Hollibaugh, J.T.; Alber, M. Watershed nitrogen input and riverine export on the west coast of the US. *Biogeochemistry* **2009**, *93*, 219–233. [CrossRef]

12. Elser, J.J.; Bracken, M.E.; Cleland, E.E.; Gruner, D.S.; Harpole, W.S.; Hillebrand, H.; Ngai, J.T.; Seabloom, E.W.; Shurin, J.B.; Smith, J.E. Global analysis of nitrogen and phosphorus limitation of primary producers in freshwater, marine and terrestrial ecosystems. *Ecol. Lett.* **2007**, *10*, 1135–1142. [CrossRef] [PubMed]

13. Pierik, M.; Van Ruijven, J.; Bezemer, T.M.; Geerts, R.H.; Berendse, F. Recovery of plant species richness during long-term fertilization of a species-rich grassland. *Ecology* **2011**, *92*, 1393–1398. [CrossRef] [PubMed]

14. Kabisch, N. Ecosystem service implementation and governance challenges in urban green space planning—The case of Berlin, Germany. *Land Use Policy* **2015**, *42*, 557–567. [CrossRef]

15. Escobedo, F.J.; Kroeger, T.; Wagner, J.E. Urban forests and pollution mitigation: Analyzing ecosystem services and disservices. *Environ. Pollut.* **2011**, *159*, 2078–2087. [CrossRef]

16. Cusack, D.F.; Silver, W.L.; Torn, M.S.; Burton, S.D.; Firestone, M.K. Changes in microbial community characteristics and soil organic matter with nitrogen additions in two tropical forests. *Ecology* **2011**, *92*, 621–632. [CrossRef]

17. Yang, L.; Zhang, L.; Li, Y.; Wu, S. Water-related ecosystem services provided by urban green space: A case study in Yixing City (China). *Landsc. Urban Plan.* **2015**, *136*, 40–51. [CrossRef]

18. Chiesura, A. The role of urban parks for the sustainable city. *Landsc. Urban Plan.* **2004**, *68*, 129–138. [CrossRef]

19. Balestrini, R.; Lumini, E. Focus on mycorrhizal symbioses. *Appl. Soil Ecol.* **2018**, *123*, 299–304. [CrossRef]

20. Bernier, N. Hotspots of biodiversity in the underground: A matter of humus form? *Appl. Soil Ecol.* **2018**, *123*, 305–312. [CrossRef]

21. Blouin, M. Chemical communication: An evidence for co-evolution between plants and soil organisms. *Appl. Soil Ecol.* **2018**, *123*, 409–415. [CrossRef]

22. Centenaro, G.; Hudek, C.; Zanella, A.; Crivellaro, A. Root-soil physical and biotic interactions with a focus on tree root systems: A review. *Appl. Soil Ecol.* **2018**, *123*, 318–327. [CrossRef]

23. Geisen, S.; Bonkowski, M. Methodological advances to study the diversity of soil protists and their functioning in soil food webs. *Appl. Soil Ecol.* **2018**, *123*, 328–333. [CrossRef]

24. Zhang, T.A.; Chen, H.Y.; Ruan, H. Global negative effects of nitrogen deposition on soil microbes. *ISME J.* **2018**, *12*, 1817–1825. [CrossRef]

25. Artursson, V.; Finlay, R.D.; Jansson, J.K. Interactions between arbuscular mycorrhizal fungi and bacteria and their potential for stimulating plant growth. *Environ. Microbiol.* **2006**, *8*, 1–10. [CrossRef]

26. Morris, S.J.; Blackwood, C.B. The ecology of the soil biota and their function. In *Soil Microbiology, Ecology and Biochemistry*, 4th ed.; Academic Press: San Diego, CA, USA, 2015; pp. 273–309.

27. Philippot, L.; Andersson, S.G.; Battin, T.J.; Prosser, J.I.; Schimel, J.P.; Whitman, W.B.; Hallin, S. The ecological coherence of high bacterial taxonomic ranks. *Nat. Rev. Microbiol.* **2010**, *8*, 523. [CrossRef]

28. Mardis, E.R. The impact of next-generation sequencing technology on genetics. *Trends Genet.* **2008**, *24*, 133–141. [CrossRef]

29. Handelsman, J. Metagenomics: Application of genomics to uncultured microorganisms. *Microbiol. Mol. Biol. Rev.* **2004**, *68*, 669–685. [CrossRef]

30. Handelsman, J.; Rondon, M.R.; Brady, S.F.; Clardy, J.; Goodman, R.M. Molecular biological access to the chemistry of unknown soil microbes: A new frontier for natural products. *Chem. Biol.* **1998**, *5*, R245–R249. [CrossRef]

31. He, Z.; Xu, M.; Deng, Y.; Kang, S.; Kellogg, L.; Wu, L.; Van Nostrand, J.D.; Hobbie, S.E.; Reich, P.B.; Zhou, J. Metagenomic analysis reveals a marked divergence in the structure of belowground microbial communities at elevated CO_2. *Ecol. Lett.* **2010**, *13*, 564–575. [CrossRef]

32. Monteiro, J.M.; Vollú, R.E.; Coelho, M.R.R.; Fonseca, A.; Neto, S.C.G.; Seldin, L. Bacterial communities within the rhizosphere and roots of vetiver (*Chrysopogon zizanioides* (L.) *Roberty*) sampled at different growth stages. *Eur. J. Soil Biol.* **2011**, *47*, 236–242. [CrossRef]

33. Cederlund, H.; Wessén, E.; Enwall, K.; Jones, C.M.; Juhanson, J.; Pell, M.; Philippot, L.; Hallin, S. Soil carbon quality and nitrogen fertilization structure bacterial communities with predictable responses of major bacterial phyla. *Appl. Soil Ecol.* **2014**, *84*, 62–68. [CrossRef]

34. Lv, X.; Ma, B.; Yu, J.; Chang, S.X.; Xu, J.; Li, Y.; Wang, G.; Han, G.; Bo, G.; Chu, X. Bacterial community structure and function shift along a successional series of tidal flats in the Yellow River Delta. *Sci. Rep.* **2016**, *6*, 36550. [CrossRef] [PubMed]

35. Fierer, N.; Strickland, M.S.; Liptzin, D.; Bradford, M.A.; Cleveland, C.C. Global patterns in belowground communities. *Ecol. Lett.* **2009**, *12*, 1238–1249. [CrossRef]
36. Zanardo, M.; Rosselli, R.; Meneghesso, A.; Sablok, G.; Stevanato, P.; Engel, M.; Altissimo, A.; Peserico, L.; Dezuani, V.; Concheri, G.; et al. Response of Bacterial Communities upon Application of Different Innovative Organic Fertilizers in a Greenhouse Experiment Using Low-Nutrient Soil Cultivated with *Cynodon dactylon*. *Soil Syst.* **2018**, *2*, 52. [CrossRef]
37. Huang, L.; Zhu, W.; Ren, H.; Chen, H.; Wang, J. Impact of atmospheric nitrogen deposition on soil properties and herb-layer diversity in remnant forests along an urban–rural gradient in Guangzhou, southern China. *Plant Ecol.* **2012**, *213*, 1187–1202. [CrossRef]
38. Lu, R.-K. *Soil Argrochemistry Analysis Protocoes*; China Agriculture Science Press: Beijing, China, 1999; Volume 106.
39. Wang, C.; Liu, D.; Bai, E. Decreasing soil microbial diversity is associated with decreasing microbial biomass under nitrogen addition. *Soil Biol. Biochem.* **2018**, *120*, 126–133. [CrossRef]
40. Zeng, J.; Liu, X.; Song, L.; Lin, X.; Zhang, H.; Shen, C.; Chu, H. Nitrogen fertilization directly affects soil bacterial diversity and indirectly affects bacterial community composition. *Soil Biol. Biochem.* **2016**, *92*, 41–49. [CrossRef]
41. Ding, J.; Zhu, D.; Chen, Q.L.; Zheng, F.; Wang, H.T.; Zhu, Y.G. Effects of long-term fertilization on the associated microbiota of soil collembolan. *Soil Biol. Biochem.* **2019**, *130*, 141–149. [CrossRef]
42. Ndour, N.Y.B.; Achouak, W.; Christen, R.; Heulin, T.; Brauman, A.; Chotte, J.L. Characteristics of microbial habitats in a tropical soil subject to different fallow management. *Appl. Soil Ecol.* **2008**, *38*, 51–61. [CrossRef]
43. Rousk, J.; Bååth, E.; Brookes, P.C.; Lauber, C.L.; Lozupone, C.; Caporaso, J.G.; Knight, R.; Fierer, N. Soil bacterial and fungal communities across a pH gradient in an arable soil. *ISME J.* **2010**, *4*, 1340. [CrossRef] [PubMed]
44. Treseder, K.K. Nitrogen additions and microbial biomass: A meta-analysis of ecosystem studies. *Ecol. Lett.* **2008**, *11*, 1111–1120. [CrossRef] [PubMed]
45. Serna-Chavez, H.M.; Fierer, N.; Van Bodegom, P.M. Global drivers and patterns of microbial abundance in soil. *Glob. Ecol. Biogeogr.* **2013**, *22*, 1162–1172. [CrossRef]
46. Lauber, C.L.; Hamady, M.; Knight, R.; Fierer, N. Soil pH as a predictor of soil bacterial community structure at the continental scale: A pyrosequencing-based assessment. *Appl. Environ. Microbiol.* **2009**, *75*, 5111–5120. [CrossRef] [PubMed]
47. Ramirez, K.S.; Lauber, C.L.; Knight, R.; Bradford, M.A.; Fierer, N. Consistent effects of nitrogen fertilization on soil bacterial communities in contrasting systems. *Ecology* **2010**, *91*, 3463–3470. [CrossRef] [PubMed]
48. Zhou, J.; Guan, D.; Zhou, B.; Zhao, B.; Ma, M.; Qin, J.; Jiang, X.; Chen, S.; Cao, F.; Shen, D.; et al. Influence of 34-years of fertilization on bacterial communities in an intensively cultivated black soil in northeast China. *Soil Biol. Biochem.* **2015**, *90*, 42–51. [CrossRef]
49. Xun, W.; Zhao, J.; Xue, C.; Zhang, G.; Ran, W.; Wang, B.; Shen, Q.; Zhang, R. Significant alteration of soil bacterial communities and organic carbon decomposition by different long-term fertilization management conditions of extremely low-productivity arable soil in South China. *Environ. Microbiol.* **2016**, *18*, 1907–1917. [CrossRef]
50. Wang, Q.; Jiang, X.; Guan, D.; Wei, D.; Zhao, B.; Ma, M.; Li, J. Long-term fertilization changes bacterial diversity and bacterial communities in the maize rhizosphere of Chinese Mollisols. *Appl. Soil Ecol.* **2018**, *125*, 88–96. [CrossRef]
51. Fan, H.B.; Liu, W.F.; Li, Y.Y.; Liao, Y.C.; Yuan, Y.H.; Xu, L. Tree growth and soil nutrients in response to nitrogen deposition in a subtropical Chinese fir plantation. *Acta Ecol. Sin.* **2007**, *27*, 4630–4642.
52. Wang, Q.; Wang, C.; Yu, W.; Turak, A.; Chen, D.; Huang, Y.; Ao, J.; Jiang, Y.; Huang, Z. Effects of nitrogen and phosphorus inputs on soil bacterial abundance, diversity, and community composition in Chinese Fir plantations. *Front. Microbiol.* **2018**, *9*, 1543. [CrossRef]
53. Nie, Y.; Wang, M.; Zhang, W.; Ni, Z.; Hashidoko, Y.; Shen, W. Ammonium nitrogen content is a dominant predictor of bacterial community composition in an acidic forest soil with exogenous nitrogen enrichment. *Sci. Total Environ.* **2018**, *624*, 407–415. [CrossRef] [PubMed]
54. Fierer, N.; Lauber, C.L.; Ramirez, K.S.; Zaneveld, J.; Bradford, M.A.; Knight, R. Comparative metagenomic, phylogenetic and physiological analyses of soil microbial communities across nitrogen gradients. *ISME J.* **2012**, *6*, 1007. [CrossRef] [PubMed]
55. Shen, C.; Xiong, J.; Zhang, H.; Feng, Y.; Lin, X.; Li, X.; Liang, W.; Chu, H. Soil pH drives the spatial distribution of bacterial communities along elevation on Changbai Mountain. *Soil Biol. Biochem.* **2013**, *57*, 204–211. [CrossRef]
56. Jones, R.T.; Robeson, M.S.; Lauber, C.L.; Hamady, M.; Knight, R.; Fierer, N. A comprehensive survey of soil Acidobacterial diversity using pyrosequencing and clone library analyses. *ISME J.* **2009**, *3*, 442. [CrossRef]
57. Yu, W.T.; Jiang, C.M.; Ma, Q.; Xu, Y.G.; Zou, H.; Zhang, S.C. Observation of the nitrogen deposition in the lower Liaohe River Plain, Northeast China and assessing its ecological risk. *Atmos. Res.* **2011**, *101*, 460–468. [CrossRef]
58. Campbell, B.J.; Polson, S.W.; Hanson, T.E.; Mack, M.C.; Schuur, E.A. The effect of nutrient deposition on bacterial communities in Arctic tundra soil. *Environ. Microbiol.* **2010**, *12*, 1842–1854. [CrossRef]
59. Lin, L.W.; Xiao, H.L.; Liu, T. Dynamics of wet atmospheric nitrogen deposition and the relation to acid rain in the northeast suburb of Guangzhou. *Ecol. Environ. Sci.* **2013**, *22*, 293–297.
60. Geisseler, D.; Scow, K.M. Long-term effects of mineral fertilizers on soil microorganisms—A review. *Soil Biol. Biochem.* **2014**, *75*, 54–63. [CrossRef]

61. Dai, Z.; Su, W.; Chen, H.; Barberán, A.; Zhao, H.; Yu, M.; Yu, L.; Brookes, P.; Schadt, C.W.; Chang, S.X.; et al. Long-term nitrogen fertilization decreases bacterial diversity and favors the growth of Actinobacteria and Proteobacteria in Agro-ecosystems across the globe. *Glob. Chang. Biol.* **2018**, *24*, 3452–3461. [CrossRef]
62. Báez, S.; Fargione, J.; Moore, D.I.; Collins, S.L.; Gosz, J.R. Atmospheric nitrogen deposition in the northern Chihuahuan desert: Temporal trends and potential consequences. *J. Arid Environ.* **2007**, *68*, 640–651. [CrossRef]
63. Torsvik, V.; Øvreås, L. Microbial diversity and function in soil: From genes to ecosystems. *Curr. Opin. Microbiol.* **2002**, *5*, 240–245. [CrossRef]
64. Van Der Heijden, M.G.; Bardgett, R.D.; Van Straalen, N.M. The unseen majority: Soil microbes as drivers of plant diversity and productivity in terrestrial ecosystems. *Ecol. Lett.* **2008**, *11*, 296–310. [CrossRef] [PubMed]

Article

Structure of Bacterial Communities in Phosphorus-Enriched Rhizosphere Soils

Yuwei Hu [1,2,3], Changqun Duan [1,4,*], Denggao Fu [1,4,*], Xiaoni Wu [1,4], Kai Yan [5], Eustace Fernando [6], Samantha C. Karunarathna [2], Itthayakorn Promputtha [7,8], Peter E. Mortimer [2] and Jianchu Xu [2]

[1] School of Ecology and Environmental Sciences & Yunnan Key Laboratory for Plateau Mountain Ecology and Restoration of Degraded Environments, Yunnan University, Kunming 650091, China; huyuwei@mail.kib.ac.cn (Y.H.); wuxiaoxiaoni@163.com (X.W.)

[2] CAS Key Laboratory for Plant Diversity and Biogeography of East Asia, Kunming Institute of Botany, Chinese Academy of Sciences, Kunming 650201, China; samantha@mail.kib.ac.cn (S.C.K.); peter@mail.kib.ac.cn (P.E.M.); jxu@mail.kib.ac.cn (J.X.)

[3] Center of Excellence in Fungal Research, Mae Fah Luang University, Chiang Rai 57100, Thailand

[4] Yunnan International Joint Research Center of Plateau Lake Ecological Restoration and Watershed Management, Yunnan University, Kunming 650091, China

[5] College of Resources and Environment, Yunnan Agricultural University, Kunming 650201, China; ecoyankai@126.com

[6] Department of Biological Sciences, Faculty of Applied Sciences, Rajarata University, Mihintale 50300, Sri Lanka; eustace6192@as.rjt.ac.lk

[7] Department of Biology, Faculty of Science, Chiang Mai University, Chiang Mai 50200, Thailand; ppam118@hotmail.com

[8] Center of Excellence in Bioresources for Agriculture, Industry and Medicine, Department of Biology, Faculty of Science, Chiang Mai University, Chiang Mai 50200, Thailand

* Correspondence: chqduan@ynu.edu.cn (C.D.); dgfu@ynu.edu.cn (D.F.); Tel.: +86-871-65032753 (C.D.)

Received: 29 August 2020; Accepted: 9 September 2020; Published: 14 September 2020

Abstract: Although phytoremediation is the main method for P-removal and maintaining ecosystem balance in geological phosphorus-enriched soils (PES), little is known about the structure and function of microbial communities in PES. Interactions between plants and soil microorganisms mainly occur in the rhizosphere. The aim of this work was to investigate the composition and diversity of bacterial communities found in rhizosphere soils associated with the following three dominant plant species: *Erianthus rufipilus*, *Coriaria nepalensis*, and *Pinus yunnanensis*. In addition, we compared these rhizosphere bacterial communities with those derived from bulk soils and grassland plots in PES from the Dianchi Lake basin of southwestern China. The Illumina MiSeq platform for high-throughput sequencing of 16S rRNA was used for the taxonomy and the analysis of soil bacterial communities. The results showed higher bacterial diversity and nutrient content in rhizosphere soils as compared with bulk soils. Rhizosphere bacteria were predominantly comprised of *Proteobacteria* (24.43%) and *Acidobacteria* (21.09%), followed by *Verrucomicrobia* (19.48%) and *Planctomycetes* (9.20%). A comparison of rhizosphere soils of the selected plant species in our study and the grassland plots showed that *Acidobacteria* were the most abundant in the rhizosphere soil of *E. rufipilus*; *Bradyrhizobiaceae* and *Rhizobiaceae* in the order *Rhizobiales* from *C. nepalensis* were found to have the greatest abundance; and *Verrucomicrobia* and *Planctomycetes* were in higher abundance in *P. yunnanensis* rhizosphere soils and in grassland plots. A redundancy analysis revealed that bacterial abundance and diversity were mainly influenced by soil water content, soil organic matter, and total nitrogen.

Keywords: phosphorus-enriched rhizosphere soils; phosphate; phytoremediation; bacterial communities; high-throughput sequencing

---

1. Introduction

Soil microbial communities are responsible for many ecosystem functions, including decomposition, nutrient cycling, nitrogen fixation, and soil formation [1–3]. Microorganisms that inhabit different soil types exhibit astonishing diversity and wide genetic variability even within species, particularly with respect to their metabolic pathways and host-interactive capabilities [4]. The majority of microbial populations in soils are concentrated in nutrient-rich niches such as rhizospheres that offer a constant supply of easily utilizable nutrients [5,6]. Recent advances in soil ecosystem microbial ecology have highlighted the interactions that occur among distinct microbial groups, now known as the "microbiome" [7].

Soil microbial communities are influenced by a series of biotic and abiotic factors [8], and specific microbial groups show high levels of diversity in different soil ecosystems [9,10]. Exogenous nutritional inputs can change structures of soil bacterial communities, promoting plant phosphorus uptake in rhizospheres [11,12]. Moreover, plant species' richness and diversity affect phosphorus uptake in rhizospheres through interactions between plant and soil bacterial communities [13]. Rhizospheric effects influence bacterial communities in rhizospheres, and therefore the diversity of rhizosphere microbial communities differs from that of the bulk soil [14].

Several studies have reported that phosphorus is a limiting factor responsible for shaping structures of soil bacterial communities across different soil habitats [15,16]. Phosphorus-rich mountains are also vulnerable and face severe degradation due to mining and other human disturbances [17]. Phosphorus removal from phosphorus-enriched soils (PES), such as in agricultural lands, grasslands, and phosphate mining regions is a major cause of eutrophication in aquatic environments and an increasing environmental problem worldwide [18]. The Dianchi Lake basin is an important industrial and agricultural region in southwest China, home to numerous PES [19,20]. Dianchi Lake is a typical plateau lake in the basin currently undergoing severe eutrophication [21]. Previous studies regarding the restoration of PES have mainly focused on the development of dominant plant communities which is a cost-effective method [17,22].

Rhizosphere microbes play important roles for sustaining ecosystems. For example, they accelerate phytoremediation processes and enhance plant ecosystem productivity [23]. Microbes inhabiting rhizospheres help plants grow and function more effectively through bolstering water retention, disease resistance, and nutrient acquisition [24]. Many studies have also shown that rhizosphere microbes promote soil remediation and restoration. Crops have been observed to reduce salt content and increase bacterial rhizosphere diversity in coastal saline soils, improving soil quality [25]. In addition, the inoculation of plant growth-promoting bacteria can change rhizosphere bacterial communities and accelerate restoration processes in eroded desert soils [26]. Rhizosphere bacteria foster arsenic bioaccumulation by plants and then improve phytoremediation of arsenic-contaminated soils [27]. However, little is known about microbes in PES. Previous studies have investigated functional bacteria taxa in PES and have identified 377 isolates that exhibited phosphate-solubilizing potential [28], although biodiversity and distribution of bacteria still remain unclear. Understanding bacterial diversity and distribution is the first step towards harnessing functional taxa to improve phytoremediation and recalibrate restoration efforts. This work highlights the urgency of studying microbial community structures in rhizospheres to support the restoration of PES.

The main objectives of this study were to investigate divergences in structures of soil bacterial communities between rhizosphere and bulk soils of plants in PES as well as understand the environmental factors that affect bacterial communities. In previous studies on phosphorus-rich mountains, efforts have centered around understanding phytoremediation. This work reports for the first time the structure of PES microbial communities influenced by dominant plants of PES.

2. Materials and Methods

2.1. Study Area

The sampling was conducted on phosphorus-rich mountains near Dianchi Lake in central Yunnan, China (Figure 1). In this study area, the average annual temperature is 14.7 °C, and the annual temperature difference is 10–15 °C. The average annual rainfall is 985.5 mm, and the rainy season lasts from May to October, accounting for 84–90% of the annual rainfall. In this study area, the variation range of the total phosphorus concentration in the soil and the nitrogen/phosphorus ratio are 1.15–80.2 g/kg and 0.006–0.98, respectively, [29] both of which are typical in a phosphorus-rich region.

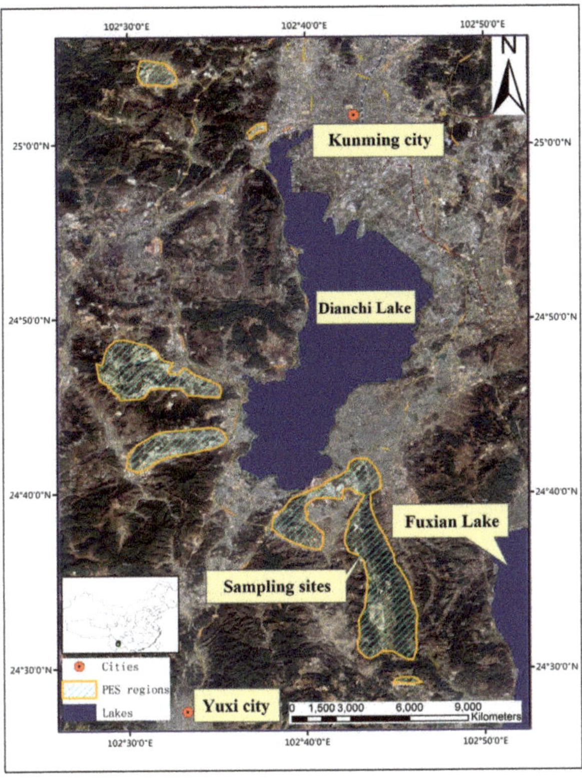

Figure 1. Geological phosphorus-rich regions around Dianchi Lake basin and sampling sites.

Most vegetation in the research area comprises natural secondary forests, and the original subtropical semi-moist evergreen broad-leaved forest has been almost completely destroyed [30]. The dominant vegetation species identified for the study were *Pinus yunnanensis*, *Coriaria nepalensis*, *Eupatorium adenophora*, *Erianthus rufipilus*, *Rumex hastatus*, *Keteleeria evelyniana*, *Pinus armandii*, and *Eucalyptus urophylla*. This study selected three dominant plants for research, i.e., *E. rufipilus*, *C. nepalensis*, and *P. yunnanensis*, based on the investigation of previous research. *E. rufipilus* belongs to the family *Gramineae*, is a drought-tolerant herbaceous plant growing between elevation 1300–2400 m, has a plant density of 2610 plant/ha and a P removal potential of 22.78 kg/ha in PES. *C. nepalensis* belongs to the family *Coriariaceae*, is an actinorhizal N-fixing species growing between elevation 400–3200 m, has a plant density of 670 plant/ha and a P removal potential of 16.91 kg/ha in PES. *P. yunnanensis* belongs to the family *Pinaceae* growing between elevation 400–3500 m, has a plant density of 497 plant/ha, a P removal potential of 27.69 kg/ha in PES, and features secondary communities that suffer from disturbance and invasion [17,22].

2.2. Soil Sampling

Fieldwork was conducted in March 2016. This study was conducted during the dry season, when rhizosphere soil acquisition is feasible, while also limiting the occurrence of rain erosion on bacterial communities. A total of thirteen sites were selected for the study area. Sampling sites were selected according to target plant communities. The soil samples in thirteen sites (Table 1) included the following: (1) rhizosphere and bulk soil of *E. rufipilus* from three sites (Er-R and Er-B), (2) rhizosphere and bulk soil of *C. nepalensis* from three sites (Cn-R and Cn-B), (3) rhizosphere and bulk soil of *P. yunnanensis* from three sites (Py-R and Py-B), and (4) soils from grassland plots from four sites (CK) with removal of surface grass debris at a random 0–20 cm soil depth in the absence of vegetation coverage. We selected plots of 10,000 m^2 (100 × 100 m) for each site. Five subplots (5 × 5 m, with at least one target plant in each plot) were chosen for soil sampling using an S-shaped sampling method [31] within sampling sites. Rhizosphere soils were collected from the fine roots of *E. rufipilus*, *C. nepalensis*, and *P. yunnanensis*. Bulk soil samples were taken at a depth of 0–20 cm in the same plots as rhizosphere soils, and then mixed thoroughly. Rhizosphere soils were collected by gently shaking off the bulk soil that adhered to the roots at a depth of 0–20 cm, and then mixed well. Each soil sample was sieved through a 2 mm mesh to remove plant roots and other plant materials. Sampling equipment including mesh, scalpels, and spoons was cleaned with 95% alcohol, followed by heating using an alcohol lamp for 30 s. Soil collections among samples were conducted after all sampling equipment was sterilized and cooled down. All soil samples were tagged and sealed in valve bags from fine roots at 0–20 cm depth. At the same time, the soil temperature (Tem) of each plot was measured using soil temperature instruments for five replicates. Next, all 22 soil samples were manually homogenized and divided into three parts. The first portion was preserved within sterile centrifuge tubes (2 mL) and put in liquid nitrogen for DNA extraction and molecular analysis; the second portion was kept in plastic bags for water content measurement; and the third portion was stored at 4 °C refrigerator for 24 h, and then dried at 25 °C for chemical analysis.

Table 1. Information of soil samples.

Abbreviation	Vegetation	Soil Location	Number of Samples
Er-R	*Erianthus rufipilus*	Rhizosphere	3
Er-B		Bulk	3
Cn-R	*Coriaria nepalensis*	Rhizosphere	3
Cn-B		Bulk	3
Py-R	*Pinus yunnanensis*	Rhizosphere	3
Py-B		Bulk	3
CK	Grassland plots	Bulk	4

2.3. Soil Physicochemical Analysis

Chemical properties for all the 22 soil samples from the study area were analyzed based on the following method: Soil water content (SWC) was determined using the oven-drying method [32]. The pH value (pH) was determined in a 1:2 (soil to water ratio) using a pH meter with a standard combination of electrodes [33]. Soil organic matter content (SOM) was determined through the potassium dichromate sulfuric acid oxidation method [34]. Total nitrogen (TN) was measured using Kjeldahl's method as modified by Bremner and Mulvaney [35]. Alkali solution nitrogen (AN) was determined using the alkali soluble diffusion method [36]. Total phosphorus (TP) content was measured using the ammonium molybdate spectrophotometric method [37]. Concentrations of available phosphorus (AP) were measured using the ammonium molybdate spectrophotometric method, following hydrochloric and sulfuric acid leaching. The value of nitrogen/phosphorus (N/P) was calculated by the ratio of total nitrogen to total phosphorus. All 22 soil samples from the four parts were used in chemical tests. Lab analyses were repeated three times, and final data were averaged.

2.4. Illumina MiSeq for High-Throughput Sequencing of 16S rRNA

Soil samples stored in liquid nitrogen were used for DNA extraction. An Agilent High Sensitivity DNA Kit (Agilent Technologies Inc., Lexington, MA, USA) was used with 0.5 g of soil, according to the manufacturer's instructions. Extraction protocol followed the manufacturer's instructions with recommended modifications to enhance the efficiency of cell lysis. To assess the quality and purity of DNA, crude DNA extracts (2 µL) of each sample were run on 2% agarose gel and analyzed using ultraviolet spectrophotometer (Eppendorf Corporation, Hamburg, Germany). Detection results of qualified DNA samples are shown in the Appendix A Table A1. Electrophoresis detection bands were single, indicating no degradation, and had no protein or RNA contamination.

Bacterial 16S rRNA genes were amplified using the following two primers: 520F (5-AYTGG GYDTAAAGNG-3) and 802R (5-TACNVGGGTATCTAATCC-3). PCR reactions were carried out in triplicate with 25 µL of reaction mixture, comprising 8.75 µL of sterilized ultrapure water, 5 µL of Q5 reaction buffer, 5 µL of Q5 GC high enhancer, 2 µL of 2.5 mM dNTPs, 2 µL of DNA template, 1 µL of each primer, and 0.25 µL Q5 polymerase. For the subsequent PCR cycling reaction, the following parameters were used: 98 °C and 2 min for initial denaturation, followed by 25 cycles of denaturation at 98 °C for 30 s, annealing at 50 °C for 30 s, and elongation at 72 °C for 30 s, followed by a final extension at 72 °C for 5 min. All the samples were amplified in triplicate and detected and quantified using a Quant-iT PicoGreen dsDNA Assay Kit (Invitrogen Corporation, Waltham, MA, USA). This step ensured that all detected DNA samples had obvious and single bands before the next experiments. After quantification, paired-end sequencing of amplicons from each reaction mixture was performed using the Illumina MiSeq platform [38]. Sequencing services were provided by Personal Biotechnology Co., Ltd. Shanghai, China. The raw sequencing data of bacteria were deposited in NCBI Sequence Read Archive (SRA) under accession number PRJNA662457.

2.5. Data Analysis

At least three groups of repeats and 22 soil samples in total from the geological phosphorus-rich mountains were collected. There were seven sets of mixed samples (Er-R, Er-B, Cn-R, Cn-B, Py-R, Py-B and CK) (as seen in Table 1) for the comparative analysis of bacterial communities. The resulting data from each set of repeated samples were averaged. For the FASTQ format of paired-end sequences, we ensured the mean quality of the base was ≥Q20 and sequence length ≥150 bp. FLASH was used to pair and connect the screened paired-end sequences in order to obtain high quality sequences [39]. UCLUST [40] was used to align sequences and call OTUs (operational taxonomic units) using 97% as the cut-off sequence similarity, showing the rarefaction curves, then removing OTUs in which the abundance value was lower than 0.001% of the total sequenced quantity [41]. Subsequent microbial classification and analyses were conducted based on classifying the information of each OTU. For the richness and diversity index of bacterial communities, the *Chao1* index (1) and *ACE* (2) index were used to describe the richness [42]:

$$chao1 = S_{obs} + \frac{f_1(f_1 - 1)}{2(f_2 + 1)} \tag{1}$$

$$ACE = \sum_{k=11}^{S_{obs}} f_k + \frac{\sum\limits_{k=1}^{10} f_k}{1 - \frac{f_1}{\sum\limits_{k=1}^{10} k f_k}} + \frac{f_1}{1 - \frac{f_1}{\sum\limits_{k=1}^{10} k f_k}} \max\left[\frac{\sum\limits_{k=1}^{10} f_k}{1 - \frac{f_1}{\sum\limits_{k=1}^{10} k f_k}} \frac{\sum\limits_{k=1}^{10} k(k-1) f_k}{(\sum\limits_{k=1}^{10} k f_k)(\sum\limits_{k=1}^{10} k f_k - 1)} - 1, 0 \right] \tag{2}$$

where S_{obs} is the total number of species observed in a sample and f_k is the number of species each represented by k individuals in a single soil sample.

The *Shannon* index (3) [43] and *Simpson* index (4) [44] were used to describe the evenness:

$$Shannon = -\sum_{i=1}^{R} p_i \ln p_i \tag{3}$$

$$Simpson = \frac{\sum_{i=1}^{R} n_i(n_i - 1)}{N(N - 1)} \tag{4}$$

where $P_i = N_i/N$, N_i is the individual number of species (i), N is the total number of organisms of a particular species, and P_i is the proportion of i species.

Raw physicochemical data properties of soil samples were processed with Excel 2007, and the results were calculated for three (rhizosphere and bulk soil of three dominant plants) or four replicate (CK) samples and expressed by the mean ± standard deviation. Statistical significance was determined by one-way analysis of variance (ANOVA), followed by Dunnett's T3 post hoc test in SPSS ver. 19.0. To compare the structure of bacterial communities across all soil samples based on the relative abundance of bacterial phyla class, principal components analysis of soil properties using redundancy analysis (RDA) by R software package was conducted. Bacterial community composition data, for each classification level, were clustered according to their number of OTUs and taxonomic data using MEGAN [45], and a phylogenetic tree based on the sequenced result of each sample and NCBI database was constructed. Heatmap and Venn diagram were drawn by hierarchal clustering performed based on Ihaka and Gentleman [46]. GraPhlAn [47] was used to construct circular representations of the taxonomic phylogenetic tree, showing the dominant taxa of rhizosphere soils and soils from grassland plots.

3. Results

3.1. Physicochemical Properties

There were significant differences among the rhizosphere soils of *E. rufipilus* (Er-R), *C. nepalensis* (Cn-R), *P. yunnanensis* (Py-R), and grassland plots (CK) (Table 1) with respect to most physicochemical properties (Table 2). The amounts of SOM, AN, TP, and AP were significantly higher in Py-R as compared with Er-R, Cn-R, and CK. Gathered data demonstrate that rhizosphere soils of *P. yunnanensis* possess a rich nutrient content

Table 2. Physiochemical properties (mean ± SD) of rhizosphere and bulk soils from three dominant plants and grassland plots.

Soil Samples	Tem (°C)	SWC (%)	pH	SOM (g/kg)	TN (g/kg)	AN (mg/kg)	TP (g/kg)	AP (mg/kg)	N/P
Er-R	15.45 ± 0.87 a	30.70 ± 5.16 a	4.48 ± 0.28 a	52.45 ± 15.68 b	2.79 ± 0.58 ab	94.06 ± 14.62 bc	5.13 ± 0.79 b	117.34 ± 44.06 b	0.55 ± 0.14 a
Er-B	15.61 ± 1.49 a	29.27 ± 2.34 a	4.34 ± 0.02 a	36.83 ± 13.75 bc	2.22 ± 0.41 ab	65.55 ± 19.15 c	4.57 ± 0.65 b	72.96 ± 30.43 b	0.49 ± 0.10 a
Cn-R	14.61 ± 1.16 a	28.35 ± 2.98 ab	4.34 ± 0.04 a	34.21 ± 5.72 bc	2.24 ± 0.11 ab	72.58 ± 14.81 c	3.93 ± 0.22 b	64.20 ± 24.64 b	0.57 ± 0.06 a
Cn-B	15.22 ± 1.48 a	26.95 ± 3.33 ab	4.32 ± 0.09 a	27.82 ± 8.39 c	1.93 ± 0.40 b	58.43 ± 18.33 c	4.20 ± 0.22 b	59.07 ± 20.74 b	0.46 ± 0.11 a
Py-R	11.09 ± 0.22 b	33.47 ± 1.03 a	4.36 ± 0.12 a	74.66 ± 10.94 a	2.99 ± 0.43 a	166.30 ± 10.24 a	11.36 ± 1.01 a	253.10 ± 30.77 a	0.27 ± 0.05 b

Table 2. *Cont.*

Soil Samples	Tem (°C)	SWC (%)	pH	SOM (g/kg)	TN (g/kg)	AN (mg/kg)	TP (g/kg)	AP (mg/kg)	N/P
Py-B	10.88 ± 0.50 b	31.38 ± 0.95 a	4.35 ± 0.05 a	39.22 ± 1.57 bc	1.83 ± 0.41 b	115.51 ± 19.16 b	12.09 ± 1.09 a	287.92 ± 56.28 a	0.15 ± 0.04 b
CK	11.09 ± 0.35 b	22.88 ± 1.15 b	4.39 ± 0.19 a	22.26 ± 2.24 c	1.92 ± 0.21 b	67.83 ± 2.06 c	3.71 ± 0.43 b	59.22 ± 11.56 b	0.52 ± 0.06 a

Notes: Value with different letters are significantly different under Dunnett's T3 post hoc test at $p < 0.05$. Er-R, rhizosphere soil of *Erianthus rufipilus*; Er-B, bulk soil of *Erianthus rufipilus*; Cn-R, rhizosphere soil of *Coriaria nepalensis*; Cn-B, bulk soil of *Coriaria nepalensis*; Py-R, rhizosphere soil of *Pinus yunnanensis*; Py-B, bulk soil of *Pinus yunnanensis*; CK, soil from grassland plots; Tem, temperature; SWC, soil water content; SOM, soil organic matter; TN, total nitrogen; AN, alkali solution nitrogen; TP, total phosphorus; AP, available phosphorus; N/P, total nitrogen/total phosphorus.

3.2. Sequencing Results

Analyses revealed an average of 40,667 valid sequences per sample, and subsequent screening analyzed 38,072 high quality sequences (accounting for 89.3–98.05%), with a read length of 200–250 bp.

3.3. Differences in Bacterial Diversity

The rarefaction curves (Figure 2) show that Py-R had the lowest bacterial diversity, and CK were more diverse as compared with the rhizosphere soils of Er-R and Cn-R. For richness and diversity estimations based on OTUs picker data, four indices of each sample group were analyzed using QIIME (Table 3). As shown in Table 3, scores for indices were as follows: Chao1 (1595) and ACE (2036) for Py-R; Chao1 (2012) and ACE (2511) for CK; the Simpson index was 0.981 and the Shannon index 8.11 for Py-R; and the Simpson index was 0.992 and the Shannon index 8.93 for Er-R. The data show that CK had the highest richness and Er-R had the highest diversity, while Py-R had the lowest richness and diversity.

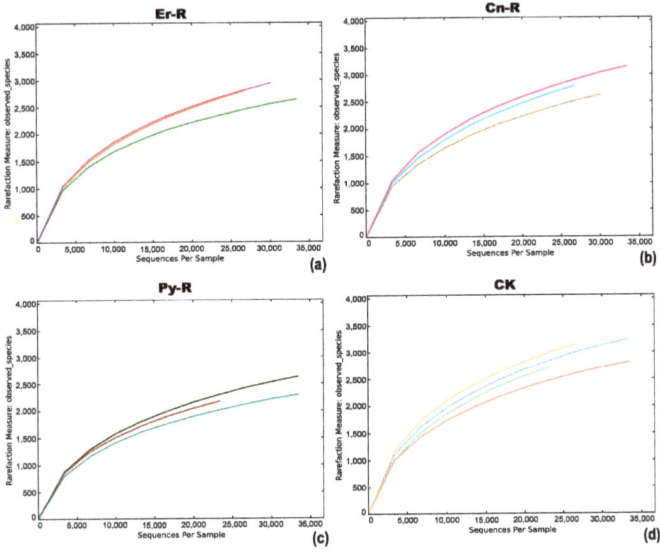

Figure 2. Rarefaction curves show the number of operational taxonomic units (OTUs) from sequencing data. Notes: Er-R, rhizosphere soil of *Erianthus rufipilus*; Cn-R, rhizosphere soil of *Coriaria nepalensis*; Py-R, rhizosphere soil of *Pinus yunnanensis*; CK, soil from grassland plots. (**a**) observed species from Er-R, (**b**) observed species from Cn-R, (**c**) observed species from Py-R, (**d**) observed species from CK.

Table 3. Comparison of richness and diversity indices among the rhizosphere soils and grassland plots.

Soil Sample	Chao1	ACE	Simpson	Shannon
Er-R	1868(1748,1953)	2302(2135,2478)	0.99151443	8.93(8.85,9.02)
Cn-R	1817(1668,2068)	2209(2010,2503)	0.989189011	8.78(8.72,8.82)
Py-R	1595(1428,1745)	2036(1736,2193)	0.981008466	8.11(7.72,8.36)
CK	2012(1682,2333)	2511(1682,3530)	0.990400468	8.91(8.42,9.24)

Notes: The values in parentheses show the upper and lower limits; Er-R, rhizosphere soil of *Erianthus rufipilus*; Cn-R, rhizosphere soil of *Coriaria nepalensis*; Py-R, rhizosphere soil of *Pinus yunnanensis*; CK, soil from grassland plots.

3.4. Bacterial Communities and Diversity

Relative bacterial OTU richness was present in four different soil samples across phylum, class, order, family, and genus (Figure 3). The dataset for phylogenetic analyses comprised 1 kingdom, 3 phyla, 7 classes, 7 orders, 16 families and 40 genera. The heatmap plot depicted the relative abundance of each dominant genera (variables clustering on the Y-axis) within soil samples (X-axis clustering) (Figure 4). As shown in Figure 4, four samples from CK grouped together; Cn-R2, Cn-R1, Er-R3, Er-R2 grouped together; and Py-R2 and Py-R3 grouped together. The 50 most abundant genera were depicted. The results show that the majority of sequences belonging to *Asteroleplasma, Enterococcus, Opitutus,* and *Alkaliphilus* were presented in soil from the CK group. *Dokdonella, Pseudomonas, Snaeromyxobacter,* and *Hyphomicrobium* were presented in soil from the Er-R and Cn-R groups. *Candidatus Xiphinematobacter, Mycobacterium, Bradyrhizobium,* and *Nostocoida* were presented in soil from the Py-R groups. We observed that the class *Proteobacteria* had the highest abundance (24.43%) in the rhizosphere soils of the three plants and soil samples from grassland plots, followed by *Acidobacteria* (21.09%), *Verrucomicrobia* (19.48%), *Planctomycetes* (9.20%), and *Anctinobacteria* (5.53%) (Figure 5).

A Venn diagram (Figure 6) illustrates that the bacterial communities of *E. rufipilus* (Er) shared 4095 OTUs between bulk (Er-B) and rhizosphere soils (Er-R) (Figure 6a); the bacterial communities of *C. nepalensis* (Cn) shared 3824 OTUs between bulk (Cn-B) and rhizosphere soils (Cn-R) (Figure 6b); the bacterial communities of *P. yunnanensis* (Py) shared 2817 OTUs between bulk (Py-B) and rhizosphere soils (Py-R) (Figure 6c); and the bacterial communities among the four groups (Er-R, Cn-R, Py-R, and CK) shared 2516 OTUs (Figure 6d). The OTUs derived from the soil associated with the three plant species showed that rhizosphere soils had more OTUs than bulk soils (Figure 6a–c).

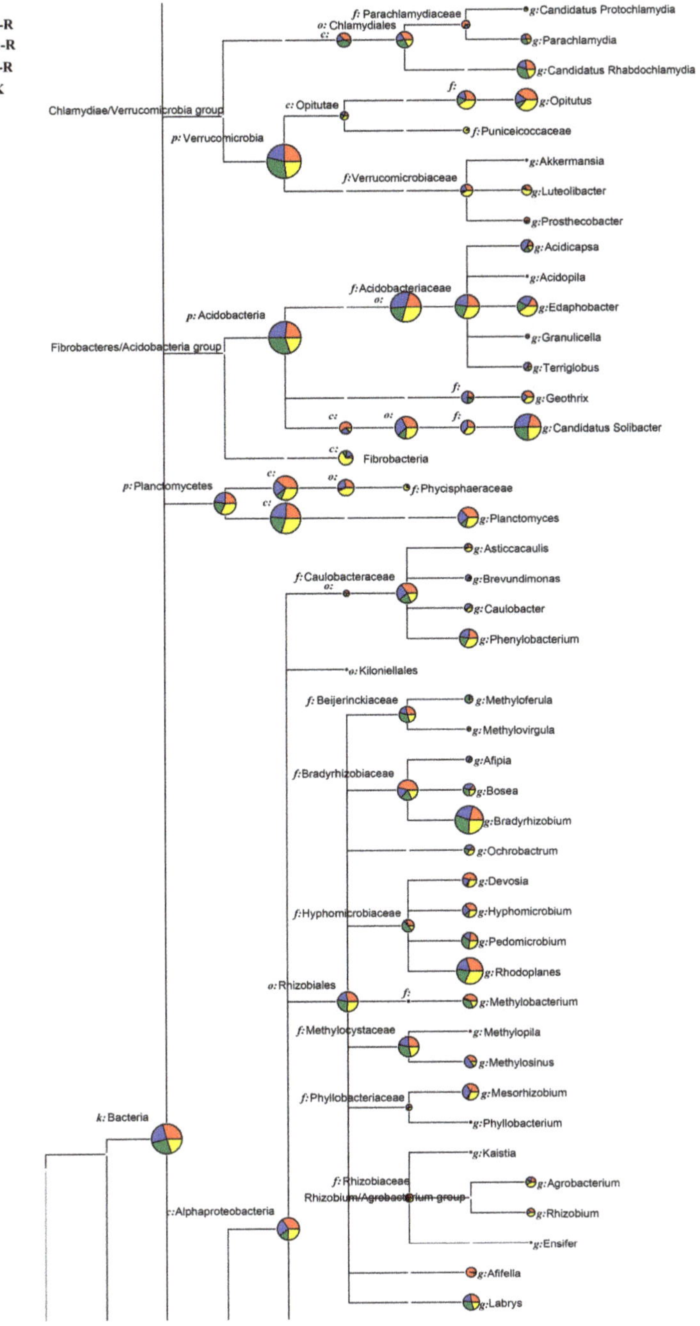

Figure 3. The phylogenetic tree of partial 16S rRNA gene read information based on NCBI Taxonomy by MEGAN. Notes: Er-R, rhizosphere soil of *Erianthus rufipilus*; Cn-R, rhizosphere soil of *Coriaria nepalensis*; Py-R, rhizosphere soil of *Pinus yunnanensis*; CK, soil from grassland plots.

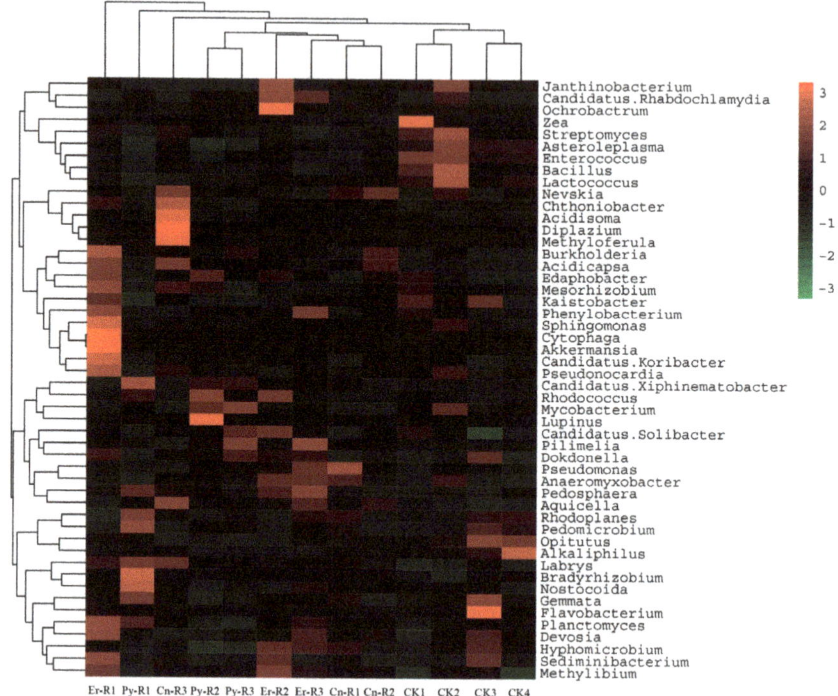

Figure 4. Hierarchically clustered heatmap of rhizosphere bacterial communities of dominant plants at the genus level. Soil samples are clustered laterally based on their bacterial similarity and bacterial taxa are clustered vertically. The red color represents the genus with higher abundance and green color represents the genus with lower abundance. Notes: Er-R, rhizosphere soil of *Erianthus rufipilus*; Cn-R, rhizosphere soil of *Coriaria nepalensis*; Py-R, rhizosphere soil of *Pinus yunnanensis*; CK, soil from grassland plots.

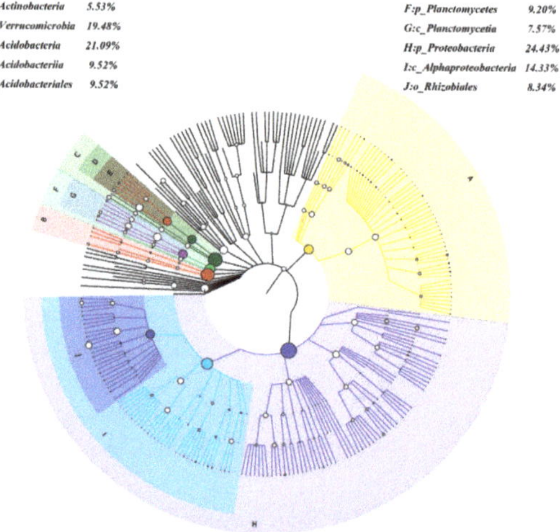

Figure 5. Hierarchical tree of dominant bacterial groups at classification level based on GraPhlAn software, it shows units of classification from phyla to genus (inner to outer rings). Size of nodes indicated the abundance.

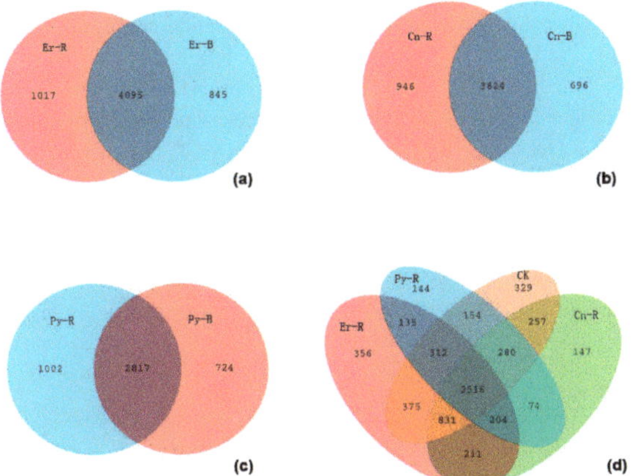

Figure 6. Numbers of shared phylotypes (OTUs) observed between soil samples of different treatments. Notes: Er-R, rhizosphere soil of *Erianthus rufipilus*; Er-B, bulk soil of *Erianthus rufipilus*; Cn-R, rhizosphere soil of *Coriaria nepalensis*; Cn-B, bulk soil of *Coriaria nepalensis*; Py-R, rhizosphere soil of *Pinus yunnanensis*; Py-B, bulk soil of *Pinus yunnanensis*; CK, soil from grassland plots. (**a**) observed OTUs in Er-R and Er-B. (**b**) observed OTUs in Cn-R and Cn-B. (**c**) observed OTUs in Py-R and Py-B. (**d**) observed OTUs in Er-R, Cn-R, Py-R and CK.

3.5. Impact of Soil Properties on the Relative Abundances of Bacterial Taxa

We found a significant association among bacterial community structure and physicochemical properties in the soils (Figure 7). Bacterial communities from Py-R, Py-B, and Cn-R tend to group together, whereas Er-R, Er-B, Cn-B, and CK are scattered. Soil water content (SWC), soil organic matter (SOM), and total nitrogen (TN) were the main limiting factors, and the first two axes in RDA analysis explained 47.24% of variance for the relationship between soil bacterial community composition and physicochemical factors.

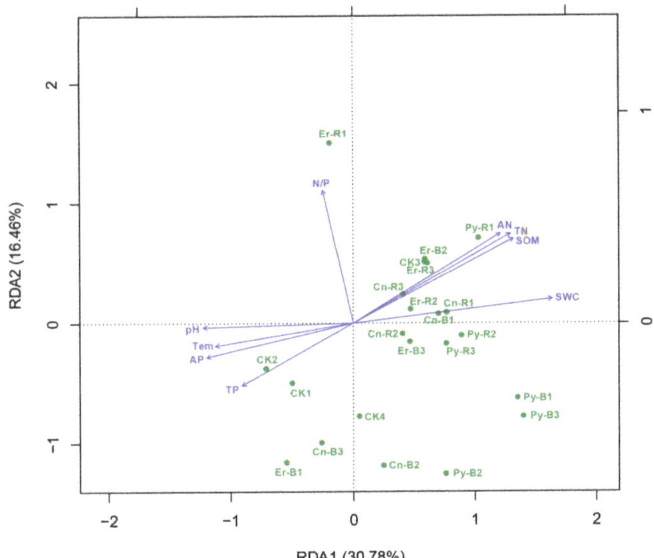

Figure 7. Redundancy analysis (RDA) of bacterial communities as affected by the environmental and

chemical properties factors, based on the relative abundance and similarity of bacterial phyla. RDA1 explained 30.78% and RDA2 16.46% variability. Each point represents a sample, the closer the distance between two points, the higher the bacterial communities' similarity. The arrow shows the influence factor, the longer the arrow, the greater the effect. The angle between the factors for acute angle shows that the two factors are positively corrected; an angle that is an obtuse angle is a negative correlation. Notes: Er-R, rhizosphere soil of *Erianthus rufipilus*; Er-B, bulk soil of *Erianthus rufipilus*; Cn-R, rhizosphere soil of *Coriaria nepalensis*; Cn-B, bulk soil of *Coriaria nepalensis*; Py-R, rhizosphere soil of *Pinus yunnanensis*; Py-B, bulk soil of *Pinus yunnanensis*; CK, soil from grassland plots; Tem, temperature; SWC, soil water content; SOM, soil organic matter; TN, total nitrogen; AN, alkali solution nitrogen; TP, total phosphorus; AP, available phosphorus; N/P, total nitrogen/total phosphorus.

4. Discussion

Results of this study demonstrate the relationships among soil physicochemical properties and the bacterial communities associated with three selected plants (*E. rufipilus*, *C. nepalensis*, and *P. yunnanensis*), as well as grassland plots in phosphorus-rich mountains in southwestern China. The associations among soil nutrients and bacterial community structure of rhizosphere and bulk soils has been demonstrated in several previous studies [48–52]. Plant species play an important role in shaping rhizosphere microbial communities through their root exudates [24,53–55]. In this study, we found that the content of SOM, TN, and AN was present in higher concentrations in rhizosphere soils as compared with bulk soils among three plants in PES. The contents of SOM, TN, AN, TP, and AP in rhizosphere soils were elevated 55.31%, 34.11%, 39.02%, −2.11%, and 3.50% relative to bulk soil, consistent with soils from other studies and regions [56,57]. The rhizosphere soils of *P. yunnanensis* have higher TP content (elevated 121.44% and 189.06% relative to *E. rufipilus* and *C. nepalensis*, respectively), AP content (elevated 115.70% and 294.24% relative to *E. rufipilus* and *C. nepalensis*, respectively), and lower nitrogen/phosphorus (N/P) (declined 50.91% and 52.63% relative to *E. rufipilus* and *C. nepalensis*, respectively) than the other two dominant plants. Differences in soil phosphorus content among land-use types in PES could possibly be due to geological reasons; for example, this study site is home to geological phosphorus-rich mountains, and phosphate deposited in Yunnan Province accounts for 80% of high-grade phosphate ore in China [17]. According to previous studies on PES, phosphorus is found in high concentrations in soils with low soil nitrogen content [21,58]. Therefore, the results of this study are congruent with previous studies.

The 16S rRNA gene sequencing survey provides exhaustive information about the relative abundance, diversity, and composition of bacterial communities. This study was the first implementation of Illumina MiSeq high-throughput sequencing technology on soil microbial communities in phosphorus-rich mountains. The relative abundance of bacteria among the soil samples showed that *P. yunnanensis* had the lowest bacterial OTUs and CK had the highest. In contrast, soil nutrient properties revealed that *P. yunnanensis* had the highest concentration of SOM, AN, TP, and AP, whereas CK had the lowest SOM. This suggests that plant species shapes soil nutrient properties and bacterial community characteristics in rhizospheres and bulk soils.

The most abundant phylum of bacterial communities in the rhizosphere soil among plants was *Proteobacteria*. This result corroborates the results of several previous bacterial community studies done on the rhizosphere soils of crops [15,59], mining soils [60,61], and PES bacterial community research [28]. Furthermore, the class *Alphaproteobacteria* occupied the majority of the phylum *Proteobacteria*. *Rhizobiales* are the most abundant order in *Proteobacteria* across all soil samples. The families *Bradyrhizobiaceae* and *Rhizobiaceae* in the order *Rhizobiales* from *C. nepalensis* have higher abundance than *E. rufipilus*, *P. yunnanensis*, and grassland plots; this is because *C. nepalensis* forms symbiotic relationships with nitrogen-fixing bacteria [62,63].

The phylum *Acidobacteria* is the second most abundant community among the rhizosphere of the plants and grassland plots in this study. *Acidobacteria* is the predominant phylum of bacteria in semi-arid and other mature soil environments [64]. *Acidobacteria* exists most abundantly in the rhizosphere soil of *E. rufipilus*; this is likely because *E. rufipilus* is highly drought resistant and suitable to grow in arid regions [65,66]. As a dominant species, *E. rufipilus* is widely distributed in phosphorus-rich regions and has a high phosphorus tolerance [63]. The next phyla in highest abundance were *Verrucomicrobia* and *Planctomycetes*. These two phyla were abundant in the rhizosphere soil of *P. yunnanensis* and grassland plots. *Planctomycetes* represented a small minority of aquatic bacteria found in seawater and freshwater environments [67,68]. However, the dry season occurs in central Yunnan, China from November to April [69], and *P. yunnanensis* exhibits drought tolerance [70], which may explain why *Planctomycetes* was present in soils linked to *P. yunnanensis* in study areas. According to the rhizosphere bacterial communities of *Y. yunnanensis* reported in other regions of Yunnan Province where phosphorus is deficient, the dominant bacterial taxa are *Acitinobacteria*, *Alphaproteobacteria*, and *Acidobacteira*, whereas *Verrucomicrobia* and *Planctomycetes* were found in low abundance [71]. In addition to the plants reported in this study, phosphorus also alters bacterial community composition in soybean rhizosphere soils, and phosphorus fertilization decreased the relative abundance of *Bacillales* and *Pseudomonadales* [72].

Plant root exudates also shape rhizosphere bacterial structures [73–75]. Previous studies about phosphorus-enriched soils have concluded that phytoremediation is an effective measure for nutrient (nitrogen and phosphorus) removal [17,76]. Fewer studies, however, has been focused on the biodiversity of soil microbial communities based on vegetation regeneration. This study showed that bacterial communities in PES are strongly influenced by the rhizospheres of different plant species and physicochemical soil parameters. The distribution of soil bacteria was affected by multiple environmental factors. Soil macro-environmental conditions, such as soil water content, has the most notable impacts on the distribution of bacteria, followed by soil organic matter and total nitrogen. Although phosphate content was high in PES, total phosphorus was found to have the lowest impact as compared with other nutrient factors on bacterial communities. The findings of this study verified the results of a previous study on low nitrogen levels in the region [19].

On a larger scale, in the phosphorus-rich and nitrogen-deficient environments of PES, most vegetation is secondary, and habitats are relatively fragile. *E. rufipilus*, *C. nepalensis*, and *P. yunnanensis* are widely grown in phosphorus-enriched soils of China because of their high tolerance to drought and arid conditions. The *P. yunnanensis* plant communities have higher plant diversity in PES, whereas bacterial communities tend to feature lower diversity under the forest soil of *P. yunnanensis*. Higher plant diversity and lower litter decomposition efficiency [77], as well as litter inhibitory effect on bacterial communities such as production of tannin [78] occurring under *P. yunnanensis* as compared with *E. rufipilus* and *C. nepalensis* likely explains this lower level of diversity. In addition, *P. yunnanensis* is a common native secondary coniferous species growing in low temperatures. Because of its rapid growth and higher plant community diversity beneath trees, it is a preferred species in Yunnan [37]. On the basis of the above findings, follow-up research in plant rhizosphere microbial compositions, and diversity could be fruitful. Nutrient properties and microbial communities can provide guidance for understanding biochemical processes and ecosystem functionalities in the soils of phosphorus-rich mountains.

5. Conclusions

The richness and diversity of the bacterial communities reported in our study reveal prospective uncultured bacteria derived from rhizosphere soils of dominant plants in PES for the first time. Differences in the physicochemistry and structures of soil bacterial communities among the dominant plants form useful guidelines for examining how tree species and soil physicochemical parameters influence local soil environments in phosphorus-rich regions. The results from our study show that both bacterial communities and soil nutrients show Rhizospheric effects; rhizosphere bacteria are mainly composed of *Proteobacteria* and *Acidobacteria*, followed by *Verrucomicrobia* and *Planctomycetes*; bacteria abundance and diversity in PES are mainly influenced by soil water content, soil organic matter, and total nitrogen. Our results indicate that rhizosphere microbial communities could serve as an important index for phytoremediation. This is of great significance for the restoration of disturbed ecosystems in PES.

Author Contributions: Conceptualization, C.D. and D.F.; Formal analysis, Y.H. and E.F.; Funding acquisition, C.D. and D.F.; Investigation, X.W. and K.Y.; Methodology, Y.H., C.D., and D.F.; Resources, C.D. and D.F.; Software, K.Y. and S.C.K.; Supervision, C.D., P.E.M., and J.X.; Writing—original draft, Y.H. and D.F. Writing—review and editing, I.P., P.E.M., and J.X. All authors have read and agree to the published version of the manuscript.

Funding: This research is supported by the National Science Foundation of China (NSFC) project codes 31670522 to Changqun Duan, NSFC project codes 31860133 to Denggao Fu, NSFC41807524 and 31861143002 to Kai Yan. Peter E. Mortimer thanks NSFC project codes 41761144055 and 41771063 for support. Samantha C. Kaunarathna thanks CAS President's International Fellowship Initiative (PIFI) for funding his postdoctoral research (no. 2018PC0006) and NSFC project code 31851110759.

Conflicts of Interest: The authors declare no conflict of interest.

Appendix A

Table A1. Basic information of UV spectrophotometry test on DNA extraction products among soil samples.

No	Samples	DNA Concentration (ng/μL)	260/280 Value	260/230 Value	Volume (μL)	Total Amount (μg)
1	Er-R1	13.84	1.80	0.56	50.00	0.69
2	Er-R2	23.54	1.65	0.62	50.00	1.18
3	Er-R3	22.43	1.79	0.65	50.00	1.12
4	Er-B1	12.87	1.92	0.53	50.00	0.64
5	Er-B2	26.76	1.76	0.74	50.00	1.34
6	Er-B3	11.39	1.64	0.45	50.00	0.57
7	Cn-R1	12.26	1.72	0.43	50.00	0.61
8	Cn-R2	14.58	1.69	0.58	50.00	0.73
9	Cn-R3	24.89	1.82	0.73	50.00	1.24
10	Cn-B1	11.73	1.69	0.46	50.00	0.59
11	Cn-B2	5.57	1.71	0.24	50.00	0.28
12	Cn-B3	5.75	1.53	0.25	50.00	0.29
13	Py-R1	15.96	1.52	0.51	50.00	0.80
14	Py-R2	16.00	1.55	0.49	50.00	0.80
15	Py-R3	17.73	1.64	0.55	50.00	0.89
16	Py-B1	10.41	1.51	0.40	50.00	0.52
17	Py-B2	9.87	1.45	0.35	50.00	0.49
18	Py-B3	10.92	1.54	0.40	50.00	0.55
19	CK1	5.44	1.39	0.29	50.00	0.27
20	CK2	10.35	1.79	0.48	50.00	0.52
21	CK3	9.49	1.72	0.40	50.00	0.47
22	CK4	11.41	1.71	0.38	50.00	0.57

References

1. McGuire, K.L.; Treseder, K.K. Microbial communities and their relevance for ecosystem models: Decomposition as a case study. *Soil Biol. Biochem.* **2010**, *42*, 529–535. [CrossRef]
2. Bhatti, A.A.; Haq, S.; Bhat, R.A. Actinomycetes benefaction role in soil and plant health. *Microb. Pathog.* **2017**, *111*, 458–467. [CrossRef] [PubMed]
3. Ravi, R.K.; Anusuya, S.; Balachandar, M.; Muthukumar, T. Microbial Interactions in Soil Formation and Nutrient Cycling. In *Mycorrhizosphere and Pedogenesis*; Varma, A., Choudhary, D.K., Eds.; Springer: Singapore, 2019; pp. 368–382. [CrossRef]
4. Berg, G.; Grube, M.; Schloter, M.; Smalla, K. Unraveling the plant microbiome: Looking back and future perspectives. *Front. Microbiol.* **2014**, *5*, 148. [CrossRef] [PubMed]
5. Dessaux, Y.; Hinsinger, P.; Lemanceau, P. Rhizosphere: So many achievements and even more challenges. *Plant Soil* **2009**, *321*, 1–3. [CrossRef]
6. Jacoby, R.P.; Kopriva, S. Metabolic niches in the rhizosphere microbiome: New tools and approaches to analyse metabolic mechanisms of plant–microbe nutrient exchange. *J. Exp. Bot.* **2019**, *70*, 1087–1094. [CrossRef]
7. Pii, Y.; Penn, A.; Terzano, R.; Crecchio, C.; Mimmo, T.; Cesco, S. Plant-microorganism-soil interactions influence the Fe availability in the rhizosphere of cucumber plants. *Plant Physiol. Biochem.* **2015**, *87*, 45–52. [CrossRef]
8. Fierer, N. Embracing the unknown: Disentangling the complexities of the soil microbiome. *Nat. Rev. Microbiol.* **2017**, *15*, 579. [CrossRef]
9. O'Brien, S.L.; Gibbons, S.M.; Owens, S.M.; Hampton–Marcell, J.; Johnston, E.R.; Jastrow, J.D.; Gilbert, J.A.; Meyer, F.; Antonopoulos, D.A. Spatial scale drives patterns in soil bacterial diversity. *Environ. Microbiol.* **2016**, *18*, 2039–2051. [CrossRef]
10. Zhang, Y.; Dong, S.; Gao, Q.; Liu, S.; Ganjurjav, H.; Wang, X.; Su, X.; Wu, X. Soil bacterial and fungal diversity differently correlated with soil biochemistry in alpine grassland ecosystems in response to environmental changes. *Sci. Rep.* **2017**, *7*, 43077. [CrossRef]
11. Shen, P.; Murphy, D.V.; George, S.J.; Lapis-Gaza, H.; Xu, M.; Gleeson, D.B. Increasing the size of the microbial biomass altered bacterial community structure which enhances plant phosphorus uptake. *PLoS ONE* **2016**, *11*, e0166062. [CrossRef]
12. Wang, Q.; Wang, C.; Yu, W.; Turak, A.; Chen, D.; Huang, Y.; Ao, J.; Jiang, Y.; Huang, Z. Effects of nitrogen and phosphorus inputs on soil bacterial abundance, diversity, and community composition in Chinese fir plantations. *Front. Microbiol.* **2018**, *9*, 1543. [CrossRef] [PubMed]
13. Wu, H.; Xiang, W.; Ouyang, S.; Forrester, D.I.; Zhou, B.; Chen, L.; Ge, T.; Lei, P.; Chen, L.; Zeng, Y.; et al. Linkage between tree species richness and soil microbial diversity improves phosphorus bioavailability. *Funct. Ecol.* **2019**, *33*, 1549–1560. [CrossRef]
14. Philippot, L.; Raaijmakers, J.M.; Lemanceau, P.; Van Der Putten, W.H. Going back to the roots: The microbial ecology of the rhizosphere. *Nat. Rev. Microbiol.* **2013**, *11*, 789–799. [CrossRef] [PubMed]
15. Samaddar, S.; Chatterjee, P.; Truu, J.; Anandham, R.; Kim, S.; Sa, T. Long-term phosphorus limitation changes the bacterial community structure and functioning in paddy soils. *Appl. Soil Ecol.* **2019**, *134*, 111–115. [CrossRef]
16. Xie, J.; Xue, W.; Li, C.; Yan, Z.; Li, D.; Li, G.; Chen, X.; Chen, D. Water-soluble phosphorus contributes significantly to shaping the community structure of rhizospheric bacteria in rocky desertification areas. *Sci. Rep.* **2019**, *9*, 18408. [CrossRef] [PubMed]
17. Yan, K.; Ranjitkar, S.; Zhai, D.; Li, Y.; Xu, J.; Li, B.; Lu, Y. Current re-vegetation patterns and restoration issues in degraded geological phosphorus-rich mountain areas: A synthetic analysis of Central Yunnan, SW China. *Plant Divers.* **2017**, *39*, 140–148. [CrossRef]
18. Sharpley, A.N.; McDowell, R.W.; Kleinman, P.J. Phosphorus loss from land to water: Integrating agricultural and environmental management. *Plant Soil* **2001**, *237*, 287–307. [CrossRef]
19. Yan, K.; Duan, C.; Fu, D.; Li, J.; Wong, M.H.G.; Qian, L.; Tian, Y. Leaf nitrogen and phosphorus stoichiometry of plant communities in geochemically phosphorus-enriched soils in a subtropical mountainous region, SW China. *Environ. Earth Sci.* **2015**, *74*, 3867–3876. [CrossRef]

20. Yan, K.; Xu, J.-C.; Gao, W.; Li, M.-J.; Li, Y.-J.; Zhao, Z.-X.; Yuan, Z.-W.; Zhang, F.-S.; Elser, J. *Human Perturbation on Phosphorus Cycles of Lake Dianchi in Southwest China*; Yunnan Agricultural University: Kunming, China, 2020; Unpublished work.

21. Yan, K.; Yuan, Z.; Goldberg, S.; Gao, W.; Ostermann, A.; Xu, J.; Zhang, F.; Elser, J. Phosphorus mitigation remains critical in water protection: A review and meta-analysis from one of China's most eutrophicated lakes. *Sci. Total Environ.* **2019**, *689*, 1336–1347. [CrossRef]

22. Yan, K.; Wong, M.H.G.; Zhang, L.; Xiong, H.; Fu, D.; Duan, C.; Guo, Z.; Wang, C. Is Phytoextraction Efficient for Remediating Phosphorus-Enriched Soils in Mountainous Region? A Case Study of Lake Dianchi Watershed of Southwestern China. *Appl. Mech. Mater.* **2014**, *448–453*, 488–493. [CrossRef]

23. Ahkami, A.H.; White III, R.A.; Handakumbura, P.P.; Jansson, C. Rhizosphere engineering: Enhancing sustainable plant ecosystem productivity. *Rhizosphere* **2017**, *3*, 233–243. [CrossRef]

24. Olanrewaju, O.S.; Ayangbenro, A.S.; Glick, B.R.; Babalola, O.O. Plant health: Feedback effect of root exudates–rhizobiome interactions. *Appl. Microbiol.* **2019**, *103*, 1155–1166. [CrossRef] [PubMed]

25. Shao, T.; Gu, X.; Zhu, T.; Pan, X.; Zhu, Y.; Long, X.; Shao, H.; Liu, M.; Rengel, Z. Industrial crop Jerusalem artichoke restored coastal saline soil quality by reducing salt and increasing diversity of bacterial community. *Appl. Soil Ecol.* **2019**, *138*, 195–206. [CrossRef]

26. Galaviz, C.; Lopez, B.R.; de–Bashan, L.E.; Hirsch, A.M.; Maymon, M.; Bashan, Y. Root growth improvement of mesquite seedlings and bacterial rhizosphere and soil community changes are induced by inoculation with plant growth–promoting bacteria and promote restoration of eroded desert soil. *Land Degrad. Dev.* **2018**, *29*, 1453–1466. [CrossRef]

27. Mesa, V.; Navazas, A.; González–Gil, R.; González, A.; Weyens, N.; Lauga, B.; Gallego, J.L.R.; Sánchez, J.; Peláez, A.I. Use of endophytic and rhizosphere bacteria to improve phytoremediation of arsenic–contaminated industrial soils by autochthonous *Betula celtiberica*. *Appl. Environ. Microbiol.* **2017**, *83*. [CrossRef]

28. Yang, P.-X.; Ma, L.; Chen, M.-H.; Xi, J.-Q.; He, F.; Duan, C.-Q.; Mo, M.-H.; Fang, D.-H.; Duan, Y.-Q.; Yang, F.-X. Phosphate solubilizing ability and phylogenetic diversity of bacteria from P-rich soils around Dianchi Lake drainage area of China. *Pedosphere* **2012**, *22*, 707–716. [CrossRef]

29. Tang, C.Q.; Zhao, M.-H.; Li, X.-S.; Ohsawa, M.; Ou, X.-K. Secondary succession of plant communities in a subtropical mountainous region of SW China. *Ecol. Res.* **2010**, *25*, 149–161. [CrossRef]

30. Tang, C.Q.; Hou, X.; Gao, K.; Xia, T.; Duan, C.; Fu, D. Man-made versus natural forests in mid-Yunnan, southwestern China. *Mt. Res. Dev.* **2007**, *27*, 242–249. [CrossRef]

31. Zhu, A.X.; Yang, L.; Li, B.; Qin, C.; Pei, T.; Liu, B. Construction of membership functions for predictive soil mapping under fuzzy logic. *Geoderma* **2010**, *155*, 164–174. [CrossRef]

32. Jarrell, W.M.; Armstrong, D.E.; Grigal, D.F.; Kelly, E.F.; Monger, H.C.; Wedin, D.A. Soil water and temperature status. In *Standard Soil Methods for Long-Term Ecological Research*; Robertson, G.P., Coleman, D.C., Bledsoe, C.S., Sollins, P., Eds.; Oxford University Press: New York, NY, USA, 1999; Volume 2, pp. 55–73.

33. Robertson, G.P.; Sollins, P.; Ellis, B.G.; Lajtha, K. Exchangeable ions, pH, and cation exchange capacity. In *Standard Soil Methods for Long-Term Ecological Research*; Robertson, G.P., Coleman, D.C., Bledsoe, C.S., Sollins, P., Eds.; Oxford University Press: New York, NY, USA, 1999; Volume 2, pp. 106–114.

34. Nelson, D.W.; Sommers, L.E. Total carbon, organic carbon, and organic matter. In *Methods of Soil Analysis, Part 3: Chemical Methods*; Sparks, D.L., Page, A.L., Helmke, P.A., Loeppert, R.H., Eds.; John Wiley & Sons: Hoboken, NJ, USA, 1996; Volume 14, pp. 961–1010.

35. Bremner, J.M. Nitrogen—Total. In *Methods of Soil Analysis, Part 3: Chemical Methods*; Sparks, D.L., Page, A.L., Helmke, P.A., Loeppert, R.H., Eds.; John Wiley & Sons: Hoboken, NJ, USA, 1996; Volume 14, pp. 1085–1121.

36. Mulvaney, R.L.; Khan, S.A.; Stevens, W.B.; Mulvaney, C.S. Improved diffusion methods for determination of inorganic nitrogen in soil extracts and water. *Biol. Fertil. Soils* **1997**, *24*, 413–420. [CrossRef]

37. Fu, D.; Wu, X.; Duan, C.; Zhao, L.; Li, B. Different life-form plants exert different rhizosphere effects on phosphorus biogeochemistry in subtropical mountainous soils with low and high phosphorus content. *Soil Tillage Res.* **2020**, *199*, 104516. [CrossRef]

38. Fadrosh, D.W.; Ma, B.; Gajer, P.; Sengamalay, N.; Ott, S.; Brotman, R.M.; Ravel, J. An improved dual-indexing approach for multiplexed 16S rRNA gene sequencing on the Illumina MiSeq platform. *Microbiome* **2014**, *2*, 6. [CrossRef]

39. Magoč, T.; Salzberg, S.L. FLASH: Fast length adjustment of short reads to improve genome assemblies. *Bioinformatics* **2011**, *27*, 2957–2963. [CrossRef] [PubMed]

40. Edgar, R.C. Search and clustering orders of magnitude faster than BLAST. *Bioinformatics* **2010**, *26*, 2460–2461. [CrossRef]

41. Bokulich, N.A.; Subramanian, S.; Faith, J.J.; Gevers, D.; Gordon, J.I.; Knight, R.; Mills, D.A.; Caporaso, J.G. Quality-filtering vastly improves diversity estimates from Illumina amplicon sequencing. *Nat. Methods* **2013**, *10*, 57–59. [CrossRef] [PubMed]

42. Gotelli, N.J.; Colwell, R.K. Estimating species richness. In *Biological Diversity: Frontiers in Measurement and Assessment*; Magurran, A.E., McGill, B.J., Eds.; Oxford University Press: New York, NY, USA, 2011; pp. 39–54.

43. Shannon, C.E. A mathematical theory of communication. *Bell Syst. Tech. J.* **1948**, *27*, 379–423. [CrossRef]

44. Simpson, E.H. Measurement of Diversity. *Nature* **1949**, *163*, 688. [CrossRef]

45. Huson, D.H.; Mitra, S.; Ruscheweyh, H.J.; Weber, N.; Schuster, S.C. Integrative analysis of environmental sequences using MEGAN4. *Genome Res.* **2011**, *21*, 1552–1560. [CrossRef]

46. Ihaka, R.; Gentleman, R. R: A language for data analysis and graphics. *J. Comput. Graph. Stat.* **1996**, *5*, 299–314.

47. Asnicar, F.; Weingart, G.; Tickle, T.L.; Huttenhower, C.; Segata, N. Compact graphical representation of phylogenetic data and metadata with GraPhlAn. *PeerJ* **2015**, *3*, e1029. [CrossRef]

48. Dotaniya, M.L.; Meena, V.D. Rhizosphere effect on nutrient availability in soil and its uptake by plants: A review. *Proc. Natl. Acad. Sci. India Sect B Biol. Sci.* **2015**, *85*, 1–12. [CrossRef]

49. Pii, Y.; Mimmo, T.; Tomasi, N.; Terzano, R.; Cesco, S.; Crecchio, C. Microbial interactions in the rhizosphere: Beneficial influences of plant growth-promoting rhizobacteria on nutrient acquisition process. A review. *Biol. Fertil. Soils* **2015**, *51*, 403–415. [CrossRef]

50. Sun, X.; Zhou, Y.; Tan, Y.; Wu, Z.; Lu, P.; Zhang, G.; Yu, F. Restoration with pioneer plants changes soil properties and remodels the diversity and structure of bacterial communities in rhizosphere and bulk soil of copper mine tailings in Jiangxi Province, China. *Environ. Sci. Pollut. Res.* **2018**, *25*, 22106–22119. [CrossRef] [PubMed]

51. Wang, Q.; Jiang, X.; Guan, D.; Wei, D.; Zhao, B.; Ma, M.; Chen, S.; Li, L.; Cao, F.; Li, J. Long-term fertilization changes bacterial diversity and bacterial communities in the maize rhizosphere of Chinese Mollisols. *Appl. Soil Ecol.* **2017**, *125*, 88–96. [CrossRef]

52. Cheng, J.; Lee, X.; Tang, Y.; Zhang, Q. Long-term effects of biochar amendment on rhizosphere and bulk soil microbial communities in a karst region, southwest China. *Appl. Soil Ecol.* **2019**, *140*, 126–134. [CrossRef]

53. Haichar, F.E.Z.; Marol, C.; Berge, O.; Rangel-Castro, J.I.; Prosser, J.I.; Balesdent, J.; Heulin, T.; Achouak, W. Plant host habitat and root exudates shape soil bacterial community structure. *ISME J.* **2008**, *2*, 1221–1230. [CrossRef]

54. Eisenhauer, N.; Lanoue, A.; Strecker, T.; Scheu, S.; Steinauer, K.; Thakur, M.P.; Mommer, L. Root biomass and exudates link plant diversity with soil bacterial and fungal biomass. *Sci. Rep.* **2017**, *7*, 1–8. [CrossRef]

55. Zhalnina, K.; Louie, K.B.; Hao, Z.; Mansoori, N.; Da Rocha, U.N.; Shi, S.; Cho, H.; Karaoz, U.; Loqué, D.; Bowen, B.P.; et al. Dynamic root exudate chemistry and microbial substrate preferences drive patterns in rhizosphere microbial community assembly. *Nat. Microbiol.* **2018**, *3*, 470–480. [CrossRef]

56. Wang, Z.; Li, T.; Wen, X.; Liu, Y.; Han, J.; Liao, Y.; DeBruyn, J.M. Fungal communities in rhizosphere soil under conservation tillage shift in response to plant growth. *Front. Microbiol* **2017**, *8*, 1301. [CrossRef]

57. Liu, J.; Ha, V.N.; Shen, Z.; Zhu, H.; Zhao, F.; Zhao, Z. Characteristics of bulk and rhizosphere soil microbial community in an ancient Platycladus orientalis forest. *Appl. Soil Ecol.* **2018**, *132*, 91–98. [CrossRef]

58. Li, M.; Hu, Z.; Zhu, X.; Zhou, G. Risk of phosphorus leaching from phosphorus-enriched soils in the Dianchi catchment, Southwestern China. *Environ. Sci. Pollut. Res.* **2015**, *22*, 8460–8470. [CrossRef] [PubMed]

59. Mendes, L.W.; Kuramae, E.E.; Navarrete, A.A.; Van Veen, J.A.; Tsai, S.M. Taxonomical and functional microbial community selection in soybean rhizosphere. *ISME J.* **2014**, *8*, 1577–1587. [CrossRef] [PubMed]

60. Li, Y.; Chen, L.; Wen, H. Changes in the composition and diversity of bacterial communities 13 years after soil reclamation of abandoned mine land in eastern China. *Ecol. Res.* **2015**, *30*, 357–366. [CrossRef]

61. Wei, Z.; Hao, Z.; Li, X.; Guan, Z.; Cai, Y.; Liao, X. The effects of phytoremediation on soil bacterial communities in an abandoned mine site of rare earth elements. *Sci. Total Environ.* **2019**, *670*, 950–960. [CrossRef] [PubMed]

62. Bargali, K.; Tewari, A. Growth and water relation parameters in drought-stressed Coriaria nepalensis seedlings. *J. Arid Environ.* **2004**, *58*, 505–512. [CrossRef]

63. Fu, D.; He, F.; Guo, Z.; Yan, K.; Wu, X.; Duan, C. Assessment of ecological restoration function of the *Coriaria nepalensis-Erianthus rufipilus* community in the phosphorus-enriched degraded mountain area in the Lake Dianchi Watershed, Southwestern China. *Chin. J. Plant Ecol.* **2013**, *37*, 326–334. (In Chinese) [CrossRef]

64. Tarlera, S.; Jangid, K.; Ivester, A.H.; Whitman, W.B.; Williams, M.A. Microbial community succession and bacterial diversity in soils during 77 000 years of ecosystem development. *FEMS Microbiol. Ecol.* **2008**, *64*, 129–140. [CrossRef]

65. Wang, X.-H.; Yang, Q.-H.; Li, F.-S.; He, L.-L.; He, S.-C. Characterization of the chromosomal transmission of intergeneric hybrids of *Saccharum* spp. and *Erianthus fulvus* by genomic in situ hybridization. *Crop Sci.* **2010**, *50*, 1642–1648. [CrossRef]

66. Evans, D.L.; Joshi, S.V.; Wang, J. Whole chloroplast genome and gene locus phylogenies reveal the taxonomic placement and relationship of *Tripidium* (Panicoideae: Andropogoneae) to sugarcane. *BMC Evol. Biol.* **2019**, *19*, 33. [CrossRef]

67. Kohn, T.; Heuer, A.; Jogler, M.; Vollmers, J.; Boedeker, C.; Bunk, B.; Rast, P.; Borchert, D.; Glöckner, I.; Freese, H.M.; et al. *Fuerstia marisgermanicae* gen. nov., sp. nov., an unusual member of the phylum Planctomycetes from the German Wadden Sea. *Front. Microbiol.* **2016**, *7*, 2079. [CrossRef]

68. Andrei, A.Ş.; Salcher, M.M.; Mehrshad, M.; Rychtecký, P.; Znachor, P.; Ghai, R. Niche-directed evolution modulates genome architecture in freshwater Planctomycetes. *ISME J.* **2019**, *13*, 1056–1071. [CrossRef]

69. Zhang, Y.; Huang, K.; Yu, Y.; Hu, T.; Wei, J. Impact of climate change and drought regime on water footprint of crop production: The case of Lake Dianchi Basin, China. *Nat. Hazards* **2015**, *79*, 549–566. [CrossRef]

70. Ma, F.; Zhao, C.; Milne, R.; Ji, M.; Chen, L.; Liu, J. Enhanced drought-tolerance in the homoploid hybrid species Pinus densata: Implication for its habitat divergence from two progenitors. *New Phytol.* **2010**, *185*, 204–216. [CrossRef] [PubMed]

71. Fan, Z.; Lu, S.; Liu, S.; Guo, H.; Wang, T.; Zhou, J.; Peng, X. Changes in Plant Rhizosphere Microbial Communities under Different Vegetation Restoration Patterns in Karst and Non–karst Ecosystems. *Sci. Rep.* **2019**, *9*, 1–12. [CrossRef]

72. Tian, J.; Lu, X.; Chen, Q.; Kuang, X.; Liang, C.; Deng, L.; Lin, D.; Cai, K.; Tian, J. Phosphorus fertilization affects soybean rhizosphere phosphorus dynamics and the bacterial community in karst soils. *Plant Soil* **2020**, 1–16. [CrossRef]

73. Cortois, R.; De Deyn, G.B. The curse of the black box. *Plant Soil* **2012**, *350*, 27–33. [CrossRef]

74. Weidner, S.; Koller, R.; Latz, E.; Kowalchuk, G.; Bonkowski, M.; Scheu, S.; Jousset, A. Bacterial diversity amplifies nutrient–based plant–soil feedbacks. *Funct. Ecol.* **2015**, *29*, 1341–1349. [CrossRef]

75. Emmett, B.D.; Buckley, D.H.; Drinkwater, L.E. Plant growth rate and nitrogen uptake shape rhizosphere bacterial community composition and activity in an agricultural field. *New Phytol.* **2020**, *225*, 960–973. [CrossRef]

76. Xiao, G.; Li, T.; Zhang, X.; Yu, H.; Huang, H.; Gupta, D.K. Uptake and accumulation of phosphorus by dominant plant species growing in a phosphorus mining area. *J. Hazard. Mater.* **2009**, *171*, 542–550. [CrossRef]

77. Fu, D.; Wu, X.; Huang, N.; Duan, C. Effects of the invasive herb *Ageratina adenophora* on understory plant communities and tree seedling growth in *Pinus yunnanensis* forests in Yunnan, China. *J. For. Res.* **2018**, *23*, 112–119. [CrossRef]

78. De Marco, A.; Esposito, F.; Berg, B.; Zarrelli, A.; Virzo De Santo, A. Litter Inhibitory Effects on Soil Microbial Biomass, Activity, and Catabolic Diversity in Two Paired Stands of *Robinia pseudoacacia* L. and *Pinus nigra* Arn. *Forests* **2018**, *9*, 766. [CrossRef]

Communication

Long-Term Exposure to Azo Dyes from Textile Wastewater Causes the Abundance of *Saccharibacteria* Population

Ramasamy Krishnamoorthy [1,†], Aritra Roy Choudhury [2,†], Polpass Arul Jose [1], Kathirvel Suganya [1,3], Murugaiyan Senthilkumar [4], James Prabhakaran [5], Nellaiappan Olaganathan Gopal [4], Jeongyun Choi [2], Kiyoon Kim [2,‡], Rangasamy Anandham [1,4,*] and Tongmin Sa [2,*]

[1] Department of Agricultural Microbiology, Agricultural College and Research Institute, Tamil Nadu Agricultural University, Madurai 625104, Tamil Nadu, India; moorthy.micro@gmail.com (R.K.); arulmku@gmail.com (P.A.J.); ksuganyaphd@gmail.com (K.S.)

[2] Department of Environmental and Biological Chemistry, Chungbuk National University, Cheongju 28644, Chungbuk, Korea; aritra.roychoudhury@outlook.com (A.R.C.); jychoi@chungbuk.ac.kr (J.C.); rldbsl67@naver.com (K.K.)

[3] Department of Environmental Science, Tamil Nadu Agricultural University, Coimbatore 641003, Tamil Nadu, India

[4] Department of Agricultural Microbiology, Tamil Nadu Agricultural University, Coimbatore 641003, Tamil Nadu, India; msenthilkumar@tnau.ac.in (M.S.); nogopal@tnau.ac.in (N.O.G.)

[5] Department of Soils and Environment, Agricultural College and Research Institute, Tamil Nadu Agricultural University, Madurai 625104, Tamil Nadu, India; prabhaharanj@gmail.com

* Correspondence: anandhamranga@gmail.com (R.A.); tomsa@chungbuk.ac.kr (T.S.)

† These authors have contributed equally to this work.

‡ Present address: Forest Medicinal Resources Research Center, National Institute of Forest Science, Yongju 36040, Korea.

Citation: Krishnamoorthy, R.; Roy Choudhury, A.; Arul Jose, P.; Suganya, K.; Senthilkumar, M.; Prabhakaran, J.; Gopal, N.O.; Choi, J.; Kim, K.; Anandham, R.; et al. Long-Term Exposure to Azo Dyes from Textile Wastewater Causes the Abundance of *Saccharibacteria* Population. *Appl. Sci.* 2021, 11, 379. https://doi.org/10.3390/app11010379

Received: 18 November 2020
Accepted: 29 December 2020
Published: 2 January 2021

Publisher's Note: MDPI stays neutral with regard to jurisdictional claims in published maps and institutional affiliations.

Abstract: Discharge of untreated wastewater is one of the major problems in various countries. The use of azo dyes in textile industries are one of the key xenobiotic compounds which effect both soil and water ecosystems and result in drastic effect on the microbial communities. Orathupalayam dam, which is constructed over Noyyal river in Tamil Nadu, India has become a sink of wastewater from the nearby textile industries. The present study had aimed to characterize the bacterial diversity and community profiles of soil collected from the vicinity of the dam (DS) and allied agricultural field (ALS) nearby the catchment area. The soil dehydrogenase and cellulase activities were significantly lower in DS compared to ALS. Additionally, the long-term exposure to azo dye compounds resulted in higher relative abundance of *Saccharibacteria* (36.4%) which are important for degradation of azo dyes. On the other hand, the relative abundance of *Proteobacteria* (25.4%) were higher in ALS. Interestingly, the abundance of *Saccharibacteria* (15.2%) were also prominent in ALS suggesting that the azo compounds might have deposited in the agricultural field through irrigation. Hence, this study revealed the potential bacterial phyla which can be key drivers for designing viable technologies for degradation of xenobiotic dyes.

Keywords: azo dye; textile; wastewater; diversity; xenobiotics; pollutant

1. Introduction

Soil is a biologically balanced system and any variations in the soil micro-environment results in the changes in native microbial community profiles. Similarly, the accumulation of pollutants such as polycyclic aromatic hydrocarbons (PAHs) [1], petroleum compounds [2] and heavy metals [3] have drastic effect on the bacterial diversity. The driving factors which determine the abundance of the microbial community in a particular contaminated soil are the genetic variation which results in altered metabolic pathways or the selective enrichment of microbes which are able to transform the particular pollutant [4].

Rapid advent of industrialization has led to deleterious effect on the environment, mostly due to the improper discharge of industrial waste. The textile industries use a

73

large amount of synthetic dyes which results in discharge of colored wastewater [5,6]. Azo dyes account for 70% of all the commercial dyes used in textile industries across the globe [7]. These dyes have been proven to be harmful for the aquatic life forms as they result in increasing the biological oxidation demand (BOD) [8]. Additionally, these azo compounds are also reported to be highly toxic and carcinogenic in nature [7]. It is estimated that around 50% of the synthetic dyes used in the textile industries does not bind to the fabric and result in getting discharged into the environment [9]. The compounds can leach into the groundwater and the use of those water sources for farming results in contamination of agricultural fields [10]. The bioremediation of contaminant by isolation and characterization of indigenous microbial community have been studied in various cases [11,12]. The classical enrichment technique does not give the broader idea about the various microbial activities occurring in the soil. Metagenomic analysis by using high-throughput sequencing helps to overcome the limitations of conventional methods of studying microbial community dynamics in soil. Researchers have concentrated on designing numerous technologies [13,14] for treatment of textile wastewater and also studied the changes in microbial community and diversity in those reactor systems which are used for remediating such wastewater [15]. However, there have been no studies on high-throughput detection of microbial community profiles of environmental samples such as water or soil which have been severely affected by long-term exposure to improper discharge of textile wastewater rich in azo dye compounds.

Orathupalayam dam is constructed on the Noyyal river in Tamil Nadu, India. The dam has become a reservoir of azo dye compounds due to the continuous discharge of effluents from the textile industries located in the nearby area of Tiruppur. In a previous study, bacterial and fungal species were isolated from the soil samples collected from the nearby catchment area and allied agricultural fields of this dam and characterized based on their ability to degrade azo dyes in a batch scale [16]. Whereas, in this study, we aimed to characterize the bacterial community profiles of soils collected from the nearby area of Orathupalayam dam and allied agricultural fields.

2. Materials and Methods

2.1. Collection of Soil Sample

Soil samples were collected from the vicinity of Orathupalayam dam and its allied agricultural lands. The dam is located (11°06′31.56″ N; 77°32′26.33″ E) in Tiruppur District, Tamil Nadu, India (Figure 1). A total of 6 composite soil samples were collected from the vicinity of the dam and 4 composite soil samples were collected from the agricultural lands situated within 2 km radius. Samples were collected by leaving at least 40 m distance between samples and 5–20 cm depth. Both the samples were made into a composite one which resulted in two soil samples: Dam Soil (DS) and Agricultural Soil (ALS).

2.2. Soil Enzyme Activities

Dehydrogenase was measured by assaying 2,3,5-triphenyl tetrazolium chloride (TTC) as substrate and defined as production of triphenylformazan (TPF) g^{-1} soil h^{-1} at 37 °C [17]. Urease activity was measured based on the amount of ammonium released after incubation of soil samples with urea solution for 2 h at 37 °C and determined calorimetrically by indophenol reaction at 690 nm [18]. Cellulase activity was measured by determining the breakdown of carboxymethylcellulose (CMC) to glucose at 30 °C for 24 h [19]. Acid phosphomonoesterase activity was determined by measuring the concentration of p-nitrophenol (p-NP) released after incubation of soil with p-NP-linked substrates for 1 h at 37 °C [20].

2.3. Soil DNA Extraction

Total soil DNA was extracted from 0.5 g of fresh soil sample by using HiPurA Soil DNA Purification Kit (Hi Media, Mumbai, India) following the manufacturer's protocol. Quality and quantity of the extracted DNA samples were checked using agarose gel electrophoresis as well as Nanodrop ND-1000 spectrophotometer (ThermoScientific, Wilmington, NC, USA).

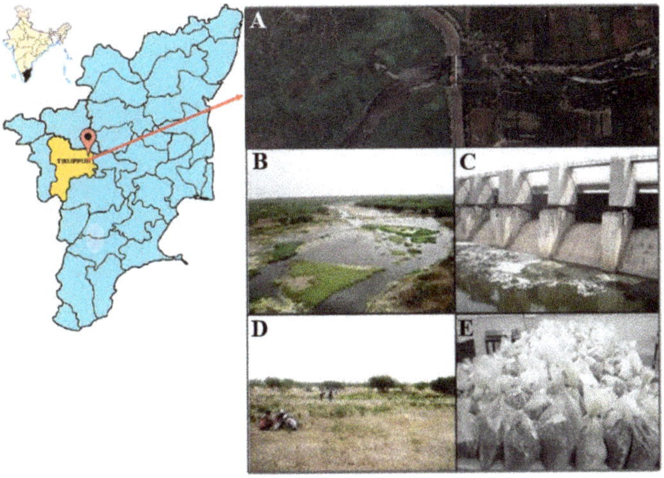

Figure 1. Overview of the soil sampling site and soil collection. (**A**) Google earth image of Orathupalayam dam constructed over Noyyal river in Tamil Nadu, India, (**B**) Orathupalayam dam inlet, (**C**) Orathupalayam dam outlet, (**D**) Soil sampling site, (**E**) Collected soil samples.

2.4. PCR Amplification and Phylogenetic Marker Library Preparation

Library construction involved two PCR reactions (Kapa HiFi Hot start, Kapa Biosystems, MA, US): amplicon PCR and indexing PCR. In amplicon PCR, the 16S rRNA gene was amplified using PCR with primers 341F, 5′-CCTACGGGAGGCAGCAG-3′ and 518R, 5′-ATTACCGCGGCTGCTGG-3′ [21–23] targeting the V3-V4 region for construction of bacterial phylogenetic marker library. Subsequently, the Illumina sequencing adapters and dual indexing barcodes were added using indexing PCR. The library of final products was cleaned using HighPrep PCR (Magbio, Gaithersburg, MD, USA) magnetic beads and quality was evaluated on a Bioanalyzer, using a DNA1000 lab chip (Agilent, USA). The raw reads were deposited in SRA archives and can be accessed by BioProject number PRJNA522349.

2.5. Pyrosequencing and Pre-Processing of Sequence Reads

Purified PCR products (library) were pooled in equimolar ratios and paired-end reads were generated on an Illumina GAIIx sequencer. Image analysis and base calling were done using Illumina Analysis pipeline (Version 2.2). High quality reads with more than 70% of bases with Phred Score greater than 20 were considered for subsequent analysis. Reads with adapter sequences, primer sequences, the barcode and the degenerate bases were removed using an automated Perl code generating processed reads. Duplicates and chimeras were removed using CD-HIT DUP [24].

2.6. Data Analysis

The resulting dataset was pre-screened using uclust for a minimum of 70% identity to ribosomal sequences and then clustered at 97% identity against the Silva (SSU115) 16S rRNA database using in MG- RAST [25]. Taxonomic assignment from phylum level to strain level was assigned based on the hits. Abundance graphs were plotted based on the number of hits. Rarefaction curves were plotted using MG-RAST [25]. Diversity index was calculated using Mothur [26]. Differences in the relative abundances of microbial community between groups were evaluated with an unequal variances t-test (Welch's *t*-test) considering unequal variance in taxonomic groups.

3. Results and Discussion

3.1. Soil Enzyme Activities

Soil quality can be determined by measuring the enzyme activities [27]. Among the various soil enzymes, dehydrogenase is considered as one of the most abundantly used parameter to determine the soil quality and reflects the rate of nutrient transformations occurring in soil. [28]. Dehydrogenase activity is affected with respect to the presence of contaminant in soil [29]. The DS had shown significantly lower dehydrogenase activity compared to ALS (Table 1). The presence of high concentration of azo compounds in DS [16] might have resulted in decrease in microbial population and subsequently reducing the enzyme activity. Results of this study match with the results of Kaczyńska et al. [28], where the authors found reduced dehydrogenase activity due to soil contamination. On the other hand, the cellulase activity is higher in soils which are rich in plant biomass [30], and it corroborates to our study where the cellulase activity has been observed to be highly significant in ALS. It can be assumed that the DS contains low amount of organic plant biomasses due to azo dye impact on plant development than that of ALS. However, urease and acid phosphomonoesterase activities did not show any significant difference among the soil samples. Urease and acid phosphomonoesterase activities are used for indirect estimation of soil nitrogen and phosphorus transformation, respectively [20].

Table 1. Soil enzyme activities of DS and ALS.

Soil Samples	Dehydrogenase (ng TPF g^{-1} Soil h^{-1})	Urease (mg NH_4^+-N g^{-1} Soil h^{-1})	Cellulase (µg Glucose g^{-1} Soil h^{-1})	Acid Phospho-monoesterase (µg Glucose g^{-1} Soil h^{-1})
DS	15.37 ± 1.96	3.36 ± 0.47 [a]	27.28 ± 1.41	54.92 ± 9.16 [a]
ALS	22.35 ± 2.68	5.11 ± 0.63 [a]	40.75 ± 4.55	55.71 ± 11.35 [a]
LSD ($p \leq 0.05$)	9.17	1.99	10.56	37.53

[a] The enzyme activities are not significantly different from each other at 5% threshold (LSD). Values are mean $\pm$ SE (standard error) of three independent determinations.

3.2. Diversity Indices

The microbial diversity is affected due to the presence of high amount of pollutants in the soil [31]. It has been observed that the azo dye contaminated soils have changes in microbial community structure based on the phospholipid fatty acid profiles [5]. Similarly, the total number of bacterial OTUs were observed to be 8302 and 12,162 for DS and ALS respectively (Table 2), after clustering at 3% cutoff level. Significant variation observed in OUT of DS and ALS, illustrated the negative impact of azo dye on soil microbial population and diversity. Similarly, the negative impact of individual azo dyes (Direct Red 81, Reactive Black 5 and Acid Yellow) on soil bacterial and fungal diversity was reported by Imran et al. [5]. In addition, the authors reported that azo dyes affect plant beneficial bacteria habitats in soil. The rarefaction curve confirmed the adequacy of sampling effort in the current microbial diversity analysis (Figure S1). The diversity indices measured in terms of Shanon and Inv-simpson indices, and the richness measured in terms of Chao1 index were higher in ALS compared to DS (Table 2). The diversity indices give the overview of richness and evenness of a community. Hence, the result of the diversity indices infers that the textile wastewater decreased the α-diversity of the microbial community and also decreased the richness and evenness of the microbial community in DS [32]. The major drawback of the study is that a single replicate was used for studying the bacterial diversity. However, it can be used as a preliminary study for understanding the effect of azo dyes on bacterial community profiles.

Table 2. Microbial diversity estimates of soil samples.

Soil Samples	Number of OTUs	Chao 1 (Richness)	Shanon (Diversity)	Inv-Simpson (Diversity)
DS	8302	1303.71	5.54	51.82
ALS	12162	1472.93	6.25	158.78

3.3. Bacterial Diversity and Community Profiles

Soil contaminated with myriad of pollutants has become a serious problem across the world. Studies which concentrate on deciphering the microbial community profiles of contaminated soil helps us to design strategies to remediate such sites by targeting the key microbial players in degradation of the particular pollutant [33]. There have been numerous studies in designing reactors for treating textile wastewater which are rich in azo compounds for decreasing the levels of effluents in the environment [13,14].

In this study, a total of 50 and 54 phyla were detected for DS and ALS respectively (Figure 2). The phyla profiles were different for both DS and ALS. *Saccharibacteria* (36.4%) were observed to be the most abundant in DS (Figure 2a) whereas *Proteobacteria* (25.4%) were the abundant ones in ALS (Figure 2b). Negative impact of azo dyes on soil *Proteobacteria* were previously reported [5]. Reactive Black 5 Direct Red 81 used by Imran et al. [5] found to reduce the *Proteobacteria* population significantly in soil. The other phyla dominating DS are *Actinobacteria* (20.02%), *Proteobacteria* (15.49%), WS5 (4.63%), *Acidobacteria* (4.16%) and *Chloroflexi* (3.31%). Similarly, the other phyla dominating ALS are *Saccharibacteria* (15.24%), *Actinobacteria* (10.03%), *Acidobacteria* (9.94%), *Plantomycetes* (8.4%), *Gemmatimonadetes* (7.59%), *Chloroflexi* (6.34%) and *Bacteroidetes* (3.48%). Different genera of *Proteobacteria* were isolated from the textile effluents wastewater treatment plants [34], and these *Proteobacteria* might have been involved in degradation of azo dyes. *Saccharibacteria* have been observed to be highly active in sludge reduction and dye wastewater treatment plants [14,35]. Furthermore, *Saccharibacteria* are regarded as one of the active microbial group in anoxic carbon-based fluidized bed reactor treating coal pyrolysis wastewater which contains abundant phenolic compounds [36]. In another study, it has been reported that *Actinobacterial* extract [37] and its related enzymes [38] are important in degradation of azo dyes. Hence, the abundance of *Saccharibacteria* and *Actinobacteria* in this study can be attributed to the fact that these phyla are responsible for remediating/decolorizing azo dye compounds present in the textile wastewater. *Saccharibacteria* and *Actinobacteria* present in Azo dye contaminated soil, indicates their important in degradation of soil pollutants. Since the results obtained by pyrosequencing techniques, many of them might be unculturable. In future, selective culturing of *Saccharibacteria* and *Actinobacteria* from this azo dye contaminated soils could be the opportunity to obtain efficient strains for remediation of azo dye polluted soils.

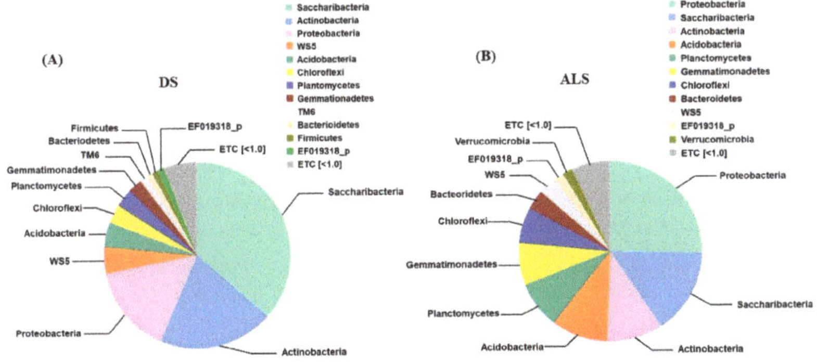

Figure 2. Relative abundance of bacterial lineages of (**A**) DS and (**B**) ALS in phylum level.

On the other hand, the abundance of *Proteobacteria* in agricultural soil also corroborates to previous studies [32]. As the agricultural sector uses mostly inorganic NPK fertilizers, which has resulted in increase in the relative abundance of *Proteobacteria* in agricultural fields [39]. The abundance of *Saccharibacteria* in agricultural soil can be correlated to the evidence that the dam water used for irrigation of the agricultural soil has resulted in contamination of the field [13,34]. The abundance of *Actinobacteria* in ALS can be explained by the fact that these bacterial group consists of plethora of plant growth promoters which are predominantly found in agricultural soils [18]. The genus and species composition of these two soil groups had revealed the presence of a number of rare bacterial species (Figure 3). There have been approximately 75% and 78% of rare bacterial species in DS and ALS, respectively. Exploiting these bacterial species can be helpful in designing efficient technologies in remediating/decolorization of azo dyes.

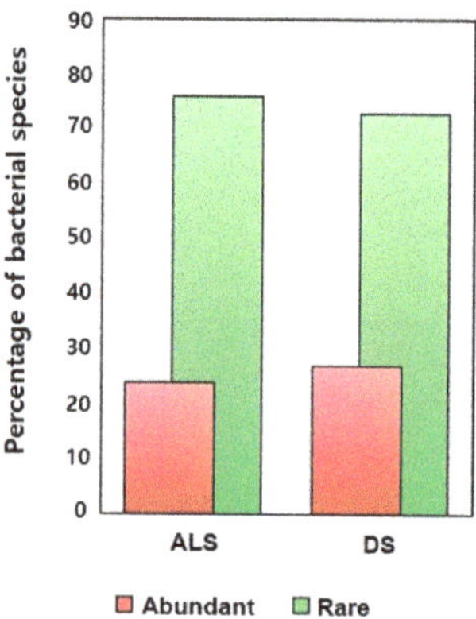

Figure 3. Percentage of abundant and rare bacterial species in ALS and DS.

4. Conclusions

The long-term exposure of wastewater from textile industries has resulted in changes in soil enzyme activities and bacterial diversity. The azo-dye contamination caused the *Saccharibacteria* population to proliferate in higher abundance compared to other bacterial phyla in DS. Moreover, ALS had shown relatively high abundance of *Saccharibacteria* which might be due to the use of polluted water from the dam for irrigation purposes. Further studies concentrating on elucidating the fungal community dynamics will help to detect the major microbial drivers important for degradation of azo dyes. The isolation and enrichment of these specific microbes (*Saccharibacteria* and *Actinobacteria*) can help to design/propose novel technologies in remediation of textile wastewater for reducing the contaminant footprint in the environment.

Supplementary Materials: The following are available online at https://www.mdpi.com/2076-3417/11/1/379/s1, Figure S1: Rarefaction curve of bacterial communities.

Author Contributions: R.A. and T.S.: conceptualization of the study; R.K. and A.R.C.: designed experiments, performed experiments and analysis of sequencing data; P.A.J., K.S., M.S., J.P. and N.O.G.: assisted in soil sampling and experiments; J.C. and K.K.: assisted in data analysis and discussion; R.K., A.R.C.: wrote the manuscript; R.A. and T.S.: critical reviewing and editing. All authors have read and agreed to the published version of the manuscript.

Funding: This work was supported by the Environmental Biotechnology task force, Department of Biotechnology (DBT), Ministry of Science and Technology, Government of India under grant no. BT/PR7518/BCE/8/994/2013 and by Basic Science Research Program, National Research Foundation (NRF), Ministry of Education, Science and Technology [2015R1A2A1A05001885], Republic of Korea.

Institutional Review Board Statement: Not applicable.

Informed Consent Statement: Not applicable.

Data Availability Statement: The data and analyses from the current study are available from the corresponding authors upon reasonable request. The raw reads were deposited in SRA archives and can be accessed by BioProject no. PRJNA522349.

Conflicts of Interest: The authors declare no conflict of interest.

References

1. Kuppusamy, S.; Thavamani, P.; Megharaj, M.; Venkateswarlu, K.; Lee, Y.B.; Naidu, R. Pyrosequencing analysis of bacterial diversity in soils contaminated long-term with PAHs and heavy metals: Implications to bioremediation. *J. Hazard. Mater.* **2016**, *317*, 169–179. [CrossRef] [PubMed]
2. Jung, J.; Philippot, L.; Park, W. Metagenomic and functional analyses of the consequences of reduction of bacterial diversity on soil functions and bioremediation in diesel-contaminated microcosms. *Sci. Rep.* **2016**, *6*, 1–10. [CrossRef] [PubMed]
3. Tang, J.; Zhang, J.; Ren, L.; Zhou, Y.; Gao, J.; Luo, L.; Yang, Y.; Peng, Q.; Huang, H.; Chen, A. Diagnosis of soil contamination using microbiological indices: A review on heavy metal pollution. *J. Environ. Manag.* **2019**, *242*, 121–130. [CrossRef] [PubMed]
4. Khan, M.A.I.; Biswas, B.; Smith, E.; Naidu, R.; Megharaj, M. Toxicity assessment of fresh and weathered petroleum hydrocarbons in contaminated soil-a review. *Chemosphere* **2018**, *212*, 755–767. [CrossRef]
5. Imran, M.; Shaharoona, B.; Crowley, D.E.; Khalid, A.; Hussain, S.; Arshad, M. The stability of textile azo dyes in soil and their impact on microbial phospholipid fatty acid profiles. *Ecotox. Environ. Safe.* **2015**, *120*, 163–168. [CrossRef]
6. Singh, L.; Singh, V.P. Textile Dyes Degradation: A Micrbial Approach for Biodegradation of Pollutants. In *Microbial Degradation of Synthetic Dyes in Wastewaters, Environmental Science and Engineering*; Singh, S.N., Ed.; Springer: Cham, Switzerland, 2015; pp. 187–204.
7. Phugare, S.S.; Kalyani, D.C.; Surwase, S.N.; Jadhav, J.P. Ecofriendly degradation, decolorization and detoxification of textile effluent by a developed bacterial consortium. *Ecotox. Environ. Safe.* **2011**, *74*, 1288–1296. [CrossRef]
8. Liu, N.; Xie, X.; Yang, B.; Zhang, Q.; Yu, C.; Zheng, X.; Xu, L.; Li, R.; Liu, J. Performance and microbial community structures of hydrolysis acidification process treating azo and anthraquinone dyes in different stages. *Environ. Sci. Pollut. Res.* **2017**, *24*, 252–263. [CrossRef]
9. Rehman, K.; Shahzad, T.; Sahar, A.; Hussain, S.; Mahmood, F.; Siddique, M.H.; Siddique, M.A.; Rashid, M.I. Effect of reactive black 5 azo dye on soil processes related to C and N cycling. *PeerJ.* **2018**, *6*, e4802. [CrossRef]
10. Zhou, Q. Chemical pollution and transport of organic dyes in water–soil–crop systems of the Chinese Coast. *Bull. Environ. Contam. Toxicol.* **2001**, *66*, 784–793.
11. Lee, D.W.; Lee, H.; Kwon, B.O.; Khim, J.S.; Yim, U.H.; Kim, B.S.; Kim, J.J. Biosurfactant assisted bioremediation of crude oil by indigenous bacteria isolated from taean beach sediment. *Environ. Pollut.* **2018**, *241*, 254–264. [CrossRef]
12. Li, Q.; Li, J.; Jiang, L.; Sun, Y.; Luo, C.; Zhang, G. Diversity and structure of phenanthrene degrading bacterial communities associated with fungal bioremediation in petroleum contaminated soil. *J. Hazard. Mater.* **2020**, *403*, 123895. [CrossRef] [PubMed]
13. Punzi, M.; Anbalagan, A.; Börner, R.A.; Svensson, B.M.; Jonstrup, M.; Mattiasson, B. Degradation of a textile azo dye using biological treatment followed by photo-Fenton oxidation: Evaluation of toxicity and microbial community structure. *Chem. Eng. J.* **2015**, *270*, 290–299. [CrossRef]
14. Xu, H.; Yang, B.; Liu, Y.; Li, F.; Song, X.; Cao, X.; Sand, W. Evolution of microbial populations and impacts of microbial activity in the anaerobic-oxic-settling-anaerobic process for simultaneous sludge reduction and dyeing wastewater treatment. *J. Clean Prod.* **2020**. [CrossRef]
15. Cui, D.; Li, G.; Zhao, D.; Gu, X.; Wang, C.; Zhao, M. Microbial community structures in mixed bacterial consortia for azo dye treatment under aerobic and anaerobic conditions. *J. Hazard. Mater.* **2012**, *221*, 185–192. [CrossRef]
16. Krishnamoorthy, R.; Jose, P.A.; Ranjith, M.; Anandham, R.; Suganya, K.; Prabhakaran, J.; Thiyageshwari, S.; Johnson, J.; Gopal, N.O.; Kumutha, K.; et al. Decolourisation and degradation of azo dyes by mixed fungal culture consisted of *Dichotomomyces cejpii* MRCH 1-2 and *Phoma tropica* MRCH 1-3. *J. Environ. Chem. Eng.* **2018**, *6*, 588–595. [CrossRef]
17. Kaczynski, P.; Lozowicka, B.; Hrynko, I.; Wolejko, E. Behaviour of mesotrione in maize and soil system and its influence on soil dehydrogenase activity. *Sci. Total Environ.* **2016**, *571*, 1079–1088. [CrossRef]
18. Samaddar, S.; Truu, J.; Chatterjee, P.; Truu, M.; Kim, K.; Kim, S.; Seshadri, S.; Sa, T. Long-term silicate fertilization increases the abundance of Actinobacterial population in paddy soils. *Biol. Fertil. Soils* **2019**, *55*, 109–120. [CrossRef]
19. Deng, S.P.; Tabatabai, M.A. Cellulase activity of soils. *Soil Biol. Biochem.* **1994**, *26*, 1347–1354. [CrossRef]
20. Colvan, S.; Syers, J.; O'Donnell, A. Effect of long-term fertiliser use on acid and alkaline phosphomonoesterase and phosphodiesterase activities in managed grassland. *Biol. Fertil. Soils* **2001**, *34*, 258–263. [CrossRef]
21. Muyzer, G.; de Waal, E.C.; Uitterlinden, A.G. Profiling of complex microbial populations by denaturing gradient gel electrophoresis analysis of polymerase chain reation-amplified genes coding for 16S rRNA. *Appl. Environ. Microbiol.* **1993**, *59*, 695–700. [CrossRef]
22. Bartram, A.K.; Lynch, M.D.; Steams, J.C.; Moreno-Hagelsieb, G.; Neufield, J.D. Generation of multimillion-sequence 16S rRNA gene libraries from complex microbial communities by assembling paired-end Illumina reads. *Appl. Environ. Microbiol.* **2011**, *77*, 3846–3852. [CrossRef] [PubMed]

23. Whiteley, A.S.; Jenkins, S.; Waite, I.; Kresoje, N.; Payne, H.; Mullan, B.; Allcock, R.; O'Donnell, A. Microbial 16S rRNA Ion Tag and community metagenome sequencing using the Ion Torrent (PGM) Platform. *J. Microbiol. Methods* **2012**, *91*, 80–88. [CrossRef] [PubMed]

24. Niu, B.; Fu, L.; Sun, S.; Li, W. Aritificial and natural duplicates in pyrosequencing reads of metagenomic data. *BMC Bioinform.* **2010**, *11*, 187. [CrossRef] [PubMed]

25. Meyer, F.; Paarmann, D.; D'Souza, M.; Olson, R.; Glass, E.M.; Kubal, M.; Paczian, T.; Rodriguez, A.; Stevens, R.; Wilke, A.; et al. The metagenomics RAST server–a public resource for the automatic phylogenetic and functional analysis of metagenomes. *BMC Bioinform.* **2008**, *9*, 1–8. [CrossRef]

26. Chao, A.; Shen, T.J. Species Prediction and Diversity Estimation (SPADE) [Software and User's Guide]. Available online: https://chao.shinyapps.io/SpadeR/http://www.chao.stat.nthu.edu.tw/softwareCE.html (accessed on 18 November 2009).

27. Badiane, N.N.Y.; Chotte, J.L.; Pate, E.; Masse, D.; Rouland, C. Use of soil enzyme activities to monitor soil quality in natural and improved fallows in semi-arid tropical regions. *Appl. Soil Ecol.* **2001**, *18*, 229–238. [CrossRef]

28. Kaczyńska, G.; Borowik, A.; Wyszkowska, J. Soil dehydrogenases as an indicator of contamination of the environment with petroleum products. *Water Air Soil Pollut.* **2015**, *226*, 372. [CrossRef]

29. Lipińska, A.; Kucharski, J.; Wyszkowska, J. The effect of polycyclic aromatic hydrocarbons on the structure of organotrophic bacteria and dehydrogenase activity in soil. *Polycycl. Aromat. Compd.* **2014**, *34*, 35–53. [CrossRef]

30. Yang, J.K.; Zhang, J.J.; Yu, H.Y.; Cheng, J.W.; Miao, L.H. Community composition and cellulase activity of cellulolytic bacteria from forest soils planted with broad-leaved deciduous and evergreen trees. *Appl. Microbiol. Biotechnol.* **2014**, *98*, 1449–1458. [CrossRef]

31. Sandaa, R.A.; Torsvik, V.; Enger, Ø. Influence of long-term heavy-metal contamination on microbial communities in soil. *Soil Biol. Biochem.* **2001**, *33*, 287–295. [CrossRef]

32. Sengupta, A.; Dick, W.A. Bacterial community diversity in soil under two tillage practices as determined by pyrosequencing. *Microb. Ecol.* **2015**, *70*, 853–859. [CrossRef]

33. Feng, G.; Xie, T.; Wang, X.; Bai, J.; Tang, L.; Zhao, H.; Wei, W.; Wang, M.; Zhao, Y. Metagenomic analysis of microbial community and function involved in cd-contaminated soil. *BMC Microbiol.* **2018**, *18*, 11. [CrossRef] [PubMed]

34. Prabha, S.; Gogoi, A.; Mazumder, P.; Ramanathan, A.L.; Kumar, M. Assessment of the impact of textile effluents on microbial diversity in Tirupur district, Tamil Nadu. *Appl. Water Sci.* **2017**, *7*, 2267–2277. [CrossRef]

35. Bai, Y.N.; Wang, X.N.; Zhang, F.; Wu, J.; Zhang, W.; Lu, Y.Z.; Fu, L.; Lau, T.C.; Zeng, R.J. High-rate anaerobic decolorization of methyl orange from synthetic azo dye wastewater in a methane-based hollow fiber membrane bioreactor. *J. Hazard. Mater.* **2020**, *388*, 121753. [CrossRef] [PubMed]

36. Zheng, M.; Shi, J.; Xu, C.; Ma, W.; Zhang, Z.; Zhu, H.; Han, H. Ecological and functional research into microbiomes for targeted phenolic removal in anoxic carbon-based fluidized bed reactor (CBFBR) treating coal pyrolysis wastewater (CPW). *Bioresour. Technol.* **2020**, *308*, 123308. [CrossRef] [PubMed]

37. Priyaragini, S.; Veena, S.; Swetha, D.; Karthik, L.; Kumar, G.; Rao, K.B. Evaluating the effectiveness of marine actinobacterial extract and its mediated titanium dioxide nanoparticles in the degradation of azo dyes. *J. Environ. Sci.* **2014**, *26*, 775–782. [CrossRef]

38. Chittal, V.; Gracias, M.; Anu, A.; Saha, P.; Rao, K.B. Biodecolorization and Biodegradation of azo dye reactive orange-16 by marine nocardiopsis sp. *Iran. J. Biotechnol.* **2019**, *17*, e1551. [CrossRef]

39. Dai, Z.; Su, W.; Chen, H.; Barberán, A.; Zhao, H.; Yu, M.; Yu, L.; Brookes, P.C.; Schadt, C.W.; Chang, S.X.; et al. Long-term nitrogen fertilization decreases bacterial diversity and favors the growth of actinobacteria and proteobacteria in agro-ecosystems across the globe. *Glob. Chang. Biol.* **2018**, *24*, 3452–3461. [CrossRef]

Article

Identification of Bacterial and Fungal Communities in the Roots of Orchids and Surrounding Soil in Heavy Metal Contaminated Area of Mining Heaps

Miroslav Böhmer [1,2,*], **Daniel Ozdín** [3], **Matúš Račko** [3], **Michal Lichvár** [4], **Jaroslav Budiš** [2,4,5] **and Tomáš Szemes** [1,2,4]

[1] Department of Molecular Biology, Faculty of Natural Sciences, Comenius University in Bratislava, Ilkovičova 6, 842 15 Bratislava, Slovakia; tomas.szemes@uniba.sk
[2] Comenius University Science Park, Ilkovičova 8, 841 04 Bratislava, Slovakia; jaroslav.budis@gmail.com
[3] Department of Mineralogy and Petrology, Faculty of Natural Sciences, Comenius University in Bratislava, Ilkovičova 6, 842 15 Bratislava, Slovakia; daniel.ozdin@uniba.sk (D.O.); matus.racko@centrum.sk (M.R.)
[4] Geneton Ltd., Ilkovičova 8, 841 04 Bratislava, Slovakia; michal.lichvar@geneton.sk
[5] Slovak Center of Scientific and Technical Information, Lamačská Cesta 8/A, 811 04 Bratislava, Slovakia
* Correspondence: bohmer6@uniba.sk

Received: 12 September 2020; Accepted: 19 October 2020; Published: 21 October 2020

Abstract: Orchids represent a unique group of plants that are well adapted to extreme conditions. In our study, we aimed to determine if different soil contamination and pH significantly change fungal and bacterial composition. We identified bacterial and fungal communities from the roots and the surrounding soil of the family *Orchidaceae* growing on different mining sites in Slovakia. These communities were detected from the samples of *Cephalanthera longifolia* and *Epipactis pontica* from Fe deposit Sirk, *E. atrorubens* from Ni-Co deposit Dobšiná and Pb-Zn deposit Jasenie and *Platanthera bifolia* by 16S rRNA gene and ITS next-generation sequencing method. A total of 171 species of fungi and 30 species of bacteria were detected from five samples of orchids. In summary, slight differences in pH of the initial soils do not significantly affect the presence of fungi and bacteria and thus the presence of the studied orchids in these localities. Similarly, the toxic elements in the studied localities, do not affect the occurrence of fungi, bacteria, and orchids. Moreover, *Cortinarius saturatus*, as a dominant fungus, and *Candidatus Udaeobacter* as a dominant bacterium were present in all soil samples and some root samples. Finally, many of these fungal and bacterial communities have the potential to be used in the bioremediation of the mining areas.

Keywords: *Orchidaceae*; soil; bacteria; fungi; microbiome; heavy metal; NGS

1. Introduction

Orchids belonging to the family *Orchidaceae* are well known for their rarity and presence of mycorrhizal associations [1,2]. This specialization contributes to the diversity and rarity of orchid species [3,4]. They grow in specific habitats such as mining dumps. Mine heaps and mine wastes created by the mining industry are one of the extreme habitats made by anthropogenic activity. Interestingly, mine heaps create an environment with specific ecological conditions for plant adaptation. They are characterized by the lack of soil as well as nutrients and moisture, and the absence of a humus layer [5]. Unique species from the family *Orchidaceae*, such as narrow-leaved Helleborine (*Cephalanthera longifolia*; Figure 1D) and Pontic Helleborine (*Epipactis pontica*; Figure 1E) which grows on the Fe deposit in Sirk, dark-red Helleborine (*E. atrorubens*; Figure 1C) from the Ni-Co deposit Dobšiná and Pb-Zn deposit Jasenie (*E. atrorubens*; Figure 1A) and Lesser Butterfly orchid (*Platanthera bifolia*; Figure 1B) from the Pb-Zn deposit Jasenie which grows on the initial soils (rankers) and tolerates high concentration of heavy metals, are well adapted to extreme conditions of the environment. These organisms germinate only after colonization with a suitable mycorrhizal fungus [6].

Figure 1. Unique species from the family *Orchidaceae*. *E. atrorubens* (**A**) and *P. bifolia* (**B**) from the Pb-Zn deposit Jasenie, *E. atrorubens* (**C**) from the Ni-Co deposit Dobšiná, *C. longifolia* (**D**) and *E. pontica* (**E**) from Fe deposit Sirk.

The fungi provide the basic physiological function for the orchids in the uptake of needed nutrients and water, otherwise inaccessible to plants [7], facilitate the emergence and survival of vegetation under stressful conditions [8], stabilize waste material through a network of veneers and improve its structure by producing substances that bind to soil particles and lead to the formation of soil aggregates [9].

The microorganisms are known to detoxify metals by several mechanisms including ion exchange, chelation, adsorption, crystallization, valence transformation, extra- and intracellular precipitation and active absorption [10]. The sorption and accumulation of metals depend on various factors such as pH, temperature, organic matter, ion species and the presence of other ions in solutions that may be in competition. They develop adaptation by temporarily altering their developmental patterns or modifying physiological characteristics depending on the toxicity of metals, which is affected by the concentration and form of salts in which the metal exists [11].

Bacteria play a key role in the element biotransformation, biogeochemical cycling, metal and mineral transformation, biological weathering and formation of the soil and sediment [12]. They have high importance in industrial processes, bioremediation, and heavy metal tolerance [13].

To identify fungi and bacterial communities the Illumina MiSeq platform was used. It is a scalable, ultra-high-throughput, cost-effective and a powerful tool to detect soil microbial communities at greater depth and with more detailed sequence information [14,15]. It has been successfully applied to identify microbial communities almost everywhere [16,17], including heavy metal contaminated environments [18,19].

The aim of this paper is to identify the composition of bacterial and fungal communities present in the roots and the surrounding soil of the orchids growing on heavy metal contaminated areas of mining heaps. To detect these communities we used 16S rRNA gene and ITS next-generation sequencing approach. At the same time, our aim was to compare the data of fungal and bacterial communities with the pH value in the Pb-Zn mining heap Jasenie-Soviansko, Ni-Co mining heap Dobšiná-Dobrá nádej and Fe mining heap Sirk-Železník. The broader goal of the study was to determine the unique microbial content of plant root collected from contaminated soil as potential biomarker applicable in further studies or monitoring efforts.

2. Materials and Methods

2.1. Site Description

Plants, soils, and geological bedrock investigated in this study originated from three abandoned mining sites in Slovakia, Central Europe (Pb-Zn deposit Jasenie-Soviansko, Ni-Co deposit Dobšiná-Dobrá nádej and Fe deposit Sirk-Železník).

The Pb-Zn deposit Jasenie-Soviansko (N 48°89′27″, E 19°44′15″) is situated in the western Ďumbier part of the Nízke Tatry Mts., 7 km north from the village Jasenie. The deposit consists of several old mining waste dumps, but orchids occurred only in the heap of the Emil adit, which is located at an altitude of 834 m a. s.l. *E. atrorubens* occurs mainly on a sharp slope of a mining heap with almost no moss or lower plants. Few orchids (*E. atrorubens* and *P. bifolia*) grew on the horizontal part of the waste dump under trees (especially spruce, willow and birch) or above heap material covered with moss, lower vegetation, and trees. *E. atrorubens* was extremely widespread in the mining heap and more than 1000 individuals grew on an area about 500 m^2. Technogenic substrate is formed mainly of metamorphic rocks of the crystalline basement of the Tatric unit of the Western Carpathians (mostly orthogneisses and migmatites) with fragments of mineralized rocks and also partially ore veins. Parts of ore veins and hydrothermally altered rocks formed hydrothermal carbonates (siderite, dolomite, ankerite, calcite), quartz and baryte and the sulphides are represented mainly as tetrahedrite, galena, bournonite, pyrite, chalcopyrite and sphalerite [20–22]. At the sampling site of the orchid and the substrate, the size of the rock fragments was up to ~7 cm. *E. atrorubens* grew in a bright habitat, without the presence of other higher plants, mosses and lichens. The slope of the mining heap with orchids was oriented to the south (SSW).

The Ni-Co deposit Dobrá nádej in Dobšiná is located in the Spišsko-Gemerské Rudohorie Mts., in the central part of the Slovak Ore Mountains Mts. in the Inner Western Carpathians (N 48°84′31″, E 20°39′08″). The mining heap is situated at an altitude of 807 m a.s.l. Orchids occur on the vertical part of a heap without vegetation but also in the horizontal part of a waste dump between low spruces and sparse low vegetation represented mainly by grasses. Samples of the orchid *E. atrorubens* and technogenic substrate were taken from the mining heap, where metamorphic rocks (phyllites, acidic metavolcanites and metasediments) dominate the Gemer superunit. Hydrothermal minerals were represented by siderite, ankerite, dolomite, quartz, hematite, Fe-Ni-Co sulphides, pyrite, chalcopyrite and minerals of tetrahedrite group [20]. Up to 4 cm was the size of the rock fragments at the sampling point of the plant and substrate. The lighting and botanical conditions at the site and the slope orientation of the mining heap were the same as in the locality Jasenie.

The Fe deposit Sirk-Železník (N 48°61′71″, E 20°10′54″) represents a weathered zone of hydrothermal Fe deposit which is located in the Revúcka vrchovina Highlands, 10 km west of Jelšava, near the village Sirk at an altitude 416 m a.s.l. Phyllites, limestones, metaryolite tuffites and metalalydites [23] occurred mostly at this deposit site. Orchids collected for research were found in the forest of a technogenic substrate with a frequent abundance of fungi and incomplete decomposed plant residues. This initial substrate was located between pieces of limonite-quartz iron ore ~8–30 cm in size. No hydrothermal carbonates were present here. Orchids grew in a light deciduous to mixed forest, mostly on a horizontal part of a mining heap between moss-covered pieces of heap material. All individuals of *E. pontica* grew in one place, in an area of ~15 m^2. *C. longifolia* grew in a disseminated forest a few meters to tens of meters from the occurrence of *E. pontica*.

Further information about localities is given in Račko et al. (2020) [20].

2.2. Sampling and DNA Extraction

The samples of the family *Orchidaceae* were taken from their place of growth. After the plant was collected, the surrounding soil was taken from this area (up to 30 cm in depth). The roots were washed with distilled water to remove remaining soil particles followed by sterilization, according to Cao et al. (2004) [24] with some modifications, with 70% ethanol for 5 min, sodium hypochlorite (2.5% available chlorine) for 5 min, and again with 70% ethanol for 1 min. After that, the roots were washed multiple times with distilled water to remove sterilization agents. Surface-sterilized roots were cut into 1–2 cm pieces and crushed on powder using mortar, pestle, and liquid nitrogen. Isolate II Plant DNA kit (Bioline, Toronto, ON, Canada) was used to extract DNA from these crushed roots. For DNA extraction from soil DNeasy PowerSoil kit (Qiagen, Hilden, Germany) was used.

2.3. Chemical Analysis

The pH of the soil was determined in the distilled H_2O and NH_4NO_3. Oxidisable organic carbon (%CO_x) was analyzed by oxidimetry under laboratory conditions. Collected soils contained various heavy metals and potentially toxic elements above limit values as described in detail at Račko et al. (2020) [20].

2.4. Polymerase Chain Reaction and Library Preparation for Next-Generation Sequencing

Bacterial 16S rRNA gene and fungal ITS region were amplified using primer pair sequences for the V3 and V4 region [25] and ITS region [26]. We used 16S Amplicon PCR Forward Primer (5′-TCG TCG GCA GCG TCA GAT GTG TAT AAG AGA CAG CCT ACG GGN GGC WGC AG-3′), 16S Amplicon PCR Reverse Primer (5′-GTC TCG TGG GCT CGG AGA TGT GTA TAA GAG ACA GGA CTA CHV GGG TAT CTA ATC C-3′), ITS1 (5′-TCG TCG GCA GCG TCA GAT GTG TAT AAG AGA CAG CTT GGT CAT TTA GAG GAA GTA A-3′) and ITS4 (5′-GTC TCG TGG GCT CGG AGA TGT GTA TAA GAG ACA GTC CTC CGC TTA TTG ATA TGC-3′). All primers contained Illumina adapter regions. PCR mixture contained HotStartTaq® Plus Master Mix (Qiagen, Hilden, Germany) (10 μl), template DNA (20 ng), and amplicon primers (2 μL, 1 μM) in the total reaction volume of 20 μl. Reaction conditions consisted of an initial 95 °C, 3 min; 35 cycles of 95 °C for 60 s, 60 °C, 45 s, and 72 °C, 60 s, and final extension of 72 °C, 7 min. After Amplicon PCR, samples were purified with DNA Clean & Concentrator™-25 (Zymo Research, Irvine, CA, USA).

We used the 16S Metagenomic Sequencing Library Preparation protocol (Illumina Inc., San Diego, CA, USA) to create libraries. Index PCR was performed following the manufacturer's instructions using 2x KAPA HiFi HotStart ReadyMix (25 μL, Kapa Biosystems, Wilmington, MA, USA), Nextera XT Index Primers (5 μL per sample, Illumina Inc., San Diego, CA, USA), PCR grade water, and 5 μL of template DNA/cDNA solution in the total reaction volume of 50 μL. PCR Clean-Up 2 was performed as described in the protocol, followed by library validation using Qubit dsDNA high sensitivity assay (LifeTechnologies, Eugene, OR, USA) and Agilent® HS DNA Chip (Agilent® Technologies 2100 Bioanalyser, Santa Clara, CA, USA). Finally, DNA libraries were normalized to 4 nM concentration, denatured and processed to the sequencing run on MiSeq instrument (Illumina Inc., San Diego, CA, USA) using a MiSeq Reagent kit v3 (Illumina Inc., San Diego, CA, USA) with the paired-end setting of 2 × 300 bp reads.

2.5. Data Analysis

Adapters and low-quality ends of sequenced reads were removed using Trimmomatic (version v0.36) [27], based on quality control statistics generated by FastQC (v0.11.5) [28]. After trimming, fragments without sufficient length of both reads (>35bp) were removed from the data set. All trimmed paired reads with sequence overlaps were merged using PEAR (v0.9.10) [29] and passed to QIIME 2 (v2019.10.0) for microbial analyses [30]. There, fastq files were dereplicated using built-in tool vsearch. Then, chimeras and "borderline chimeras" [31] were removed from the dataset using de novo chimera filtering with default parameters [30]. Operational taxonomic units (OTUs) with 99% similarity were created with de novo clustering of features.

Taxonomy of OTUs was assigned by QIIME's pre-fitted sklearn-based taxonomy classifier [32]. The pre-fitted classifier was trained by the Naive–Bayes classifier with default parameters on Silva 16S database (Silva release 132) [33] and ITS database (UNITE 8.2) [34]. Only taxons with <0.1% of all reads per sample were used for bacteria and <0.01% for fungi. Finally, Faith's phylogenetic diversity for all samples was calculated using faith-pd plugin [35]. All computational analyses were written and executed using the SnakeLines framework (v0.9.2) [36,37].

3. Results and Discussion

Orchidaceae is a family of rare plants and their presence in metal-rich biotopes has ecological and conservation importance [38]. Orchids colonizing mining heaps rely on the fungal mycelium to obtain the necessary nutrients since their root system is not well developed. They belong to metal excluders, and mycorrhizal fungi seem to play an important role in filtering out the heavy metals [38–40]. Despite these extreme conditions, the environment is sufficiently tolerable to the population of orchids, soil filamentous fungi and bacteria.

In orchidaceous mycorrhiza, fungi have the ability to support the composition of plants in stressed ecological niches and to increase the plant's condition [41]. The properties of orchidaceous mycorrhiza in mycoheterotrophic plants that allow colonization under stress conditions are changes in root morphology, orchid metabolism, such as enzymatic activities, secretion of organic acids with low molecular weight and phenols, and associations with orchideous mycorrhizal fungus adapted to the environmental conditions in which orchids occur [42,43]. To identify fungal and bacterial communities in the roots of orchids and surrounding soil in heavy metal contaminated areas we used the marker gene sequencing approach which has been the most commonly used method to analyze these communities. It is simple, fast, cost-effective, and well-tested approach [44,45] in comparison to whole-genome sequencing [45,46].

A total of 171 species of fungi were detected from five samples of orchids, belonging to seven phyla, *Ascomycota, Basidiobolomycota, Basidiomycota, Chytridiomycota, Monoblepharomycota, Mortierellomycota,* and *Mucoromycota* (Table 1). The highest biodiversity of fungi was situated in the roots of *E. pontica* (Sirk), and the surrounding soil of *E. atrorubens* (Jasenie). Moreover, the most abundant phylum of fungi was *Basidiomycota* in both soil and the roots of orchids. Furthermore, the most abundant species of fungi in the roots of orchids was *Cortinarius saturatus, Mycena* sp., *Mycena citrinomarginata, Mycenella bryophila, Colletotrichum capsici, Trichophaea pseudogregaria, Phacidium* sp., *Russula solaris, Mortierella* sp., *Amphinema byssoides, Mycena sanguinolenta, Amphinema* sp., and *Piloderma byssinum* (Figure 2). Further, in the surrounding soil, it was *Cortinarius saturatus, Cortinarius scandens, Helvellosebacina concrescens, Mortierella* sp., *Cortinarius* sp., *Helvellosebacina* sp., *Cuphophyllus virgineus, Tetracladium* sp., *Colletotrichum capsici, Cortinarius vernus, Penicillium* sp., *Saitozyma podzolica,* and *Solicoccozyma terricola* (Figure 3). In addition, we observed that the biodiversity in sampled pairs were consistently higher in soil samples, both in the number of discovered species (median 2.7x higher) and the evenness (median 2.66x higher of Faith's phylogenetic diversity). Decreased diversity could be due to the specific environment in the roots of the orchids.

Table 1. DNA identification of the fungi isolated from the roots of the orchids and surrounding soils.

The Most Abundant Strains of Fungi Detected in the Roots of Orchids and Surrounding Soils	1 *Epipactis atrorubens* Ni-Co Mining Heap Dobšiná pH 7.73		2 *Epipactis atrorubens* Pb-Zn Mining Heap Jasenie pH 7.36		3 *Platanthera bifolia* pH 7.81		4 *Epipactis pontica* Fe Mining Heap Sirk pH 4.98		5 *Cephalanthera longifolia* pH 6.58	
	Roots	Soil	Roots	Soil	Roots	Soil	Roots	Soil	Roots	Soil
Ascomycota										
Alatospora acuminata	−	−	−	+	−	−	−	−	−	−
Alatospora sp.	−	−	−	+	−	−	−	−	−	−
Arthrobotrys conoides	−	+	−	−	−	−	−	−	−	−
Arthrobotrys sp.	−	+	−	−	−	−	−	−	−	−
Aureobasidium pullulans	−	+	−	−	−	−	−	−	−	−
Bradymyces alpinus	−	+	−	−	−	−	−	−	−	−
Cadophora sp.	−	+	−	−	−	−	−	−	−	−
Capronia sp.	−	−	−	−	−	+	−	−	−	−
Cenococcum geophilum	−	−	−	−	−	−	−	−	+	+
Cenococcum sp.	−	−	−	−	−	−	+	+	+	+
Cistella sp.	−	−	−	+	−	−	−	−	−	−
Cladophialophora minutissima	−	−	−	+	−	−	−	−	−	−
Cladosporium sp.	−	+	−	−	−	−	−	−	−	−
Colletotrichum capsici	−	+	+	+	+	+	+	+	+	+
Dothidea eucalypti	−	−	−	+	−	−	−	−	−	−
Exophiala equina	−	−	−	+	−	−	+	+	−	+
Exophiala radicis	−	−	−	+	−	−	+	+	−	+
Exophiala sp.	−	−	+	+	−	−	+	+	−	+
Geopora arenicola	−	+	−	−	−	−	−	−	−	−
Geopora sp.	−	+	−	−	−	−	−	−	−	−
Geoglossum fallax	−	−	−	−	−	−	+	+	−	−
Geomyces auratus	−	−	−	−	−	+	−	−	−	−
Gyoerffyella sp.	−	−	−	−	−	−	+	−	−	−
Herpotrichia sp.	−	−	−	−	−	−	−	−	+	−
Humaria hemisphaerica	−	−	−	−	−	−	+	−	−	−
Humaria sp.	−	−	−	−	−	−	+	−	−	−
Hyalodendriella_betulae	−	−	−	+	−	−	−	−	−	−
Hyalodendriella sp.	−	−	−	+	−	−	−	−	−	−
Hyaloscypha bicolor	−	−	−	−	−	+	−	+	−	−
Hyaloscypha finlandica	−	−	−	+	−	−	−	+	+	−
Hyaloscypha sp.	−	−	−	−	−	−	−	−	+	−
Infundichalara minuta	−	−	−	+	−	−	−	−	−	−
Lachnum sp.	−	−	−	−	−	−	+	−	−	−
Leohumicola sp.	−	−	−	+	−	−	−	−	−	−
Mollisia dextrinospora	−	−	−	+	−	−	−	−	−	−
Sporormiella intermedia	−	−	−	+	−	−	−	−	−	−
Paracladophialophora carceris	−	−	−	−	−	−	−	+	−	−
Paraphoma fimeti	−	+	−	−	−	−	−	−	−	−
Penicillium brunneoconidiatum	−	−	−	+	−	+	−	−	−	−
Penicillium spinulosum	−	−	−	+	−	+	−	−	−	−
Penicillium sp.	−	−	−	+	−	+	−	+	−	+
Peziza succosa	−	−	−	+	−	−	−	+	−	−
Pezoloma ericae	−	−	+	+	−	+	−	+	+	−
Phacidium sp.	−	−	+	−	−	−	−	−	−	−
Phialocephala fortinii	−	−	−	−	+	−	−	−	−	−
Phialocephala sp.	−	−	−	−	−	+	−	−	−	−
Plenodomus biglobosus	−	+	−	−	−	−	−	−	−	−
Pseudodictyosporium elegans	−	+	−	−	−	−	−	−	−	−
Pseudodictyosporium sp.	−	+	−	−	−	−	−	−	−	−
Sagenomella sp.	−	−	−	−	−	+	−	−	−	−
Talaromyces sp.	−	−	−	+	−	+	−	−	−	−
Tetracladium apiense	−	−	+	−	−	−	−	−	−	−
Tetracladium sp.	−	+	+	+	−	−	−	−	−	−
Tricharina gilva	−	+	−	−	−	−	−	−	−	−
Tricharina sp.	−	+	−	−	−	−	−	−	−	−
Trichocladium opacum	−	−	−	−	−	−	−	+	−	−
Trichophaea pseudogregaria	+	+	−	−	−	−	−	−	−	−
Trichophaea woolhopeia	+	−	+	+	−	−	−	−	−	−
Tuber rufum	−	−	−	−	−	−	−	−	−	+
Tuber sp.	−	−	−	−	−	−	−	−	−	+
Verrucaria muralis	−	+	−	−	−	−	−	−	−	−
Verrucaria ahtii	−	+	−	−	−	−	−	−	−	−
Basidiobolomycota										
Basidiobolus magnus	−	−	−	+	−	−	−	−	−	−
Basidiobolus ranarum	−	−	−	+	−	−	−	−	−	−

Table 1. *Cont.*

The Most Abundant Strains of Fungi Detected in the Roots of Orchids and Surrounding Soils	1 *Epipactis atrorubens* Ni-Co Mining Heap Dobšiná pH 7.73		2 *Epipactis atrorubens* Pb-Zn Mining Heap Jasenie pH 7.36		3 *Platanthera bifolia* pH 7.81		4 *Epipactis pontica* Fe Mining Heap Sirk pH 4.98		5 *Cephalanthera longifolia* pH 6.58	
	Roots	Soil	Roots	Soil	Roots	Soil	Roots	Soil	Roots	Soil
Basidiomycota										
Amphinema byssoides	−	−	+	+	−	−	−	−	−	−
Amphinema sp.	−	−	+	+	−	+	−	−	−	−
Apiotrichum dulcitum	−	−	−	−	−	−	−	+	−	−
Apiotrichum sp.	−	−	−	−	−	−	−	+	−	−
Coprinellus micaceus	−	−	−	+	−	−	−	−	−	−
Cortinarius casimiri	−	−	−	−	−	+	−	−	−	−
Cortinarius cyprinus	−	−	−	−	−	−	−	−	−	+
Cortinarius saturatus	+	+	+	+	+	+	+	+	+	+
Cortinarius scandens	−	−	−	−	+	+	−	−	−	−
Cortinarius subtilis	−	−	−	+	−	−	−	−	−	−
Cortinarius vernus	−	−	−	−	−	+	+	+	−	−
Cortinarius sp.	−	−	−	+	+	+	+	+	−	+
Cuphophyllus virgineus	−	−	−	−	−	−	+	+	−	−
Cuphophyllus sp.	−	−	−	−	−	−	−	+	−	−
Cutaneotrichosporon moniliiforme	−	−	−	−	−	−	−	−	+	−
Cutaneotrichosporon sp.	−	−	−	−	−	−	+	−	−	−
Filobasidium wieringae	−	+	−	−	−	−	−	−	−	−
Flagelloscypha minutissima	−	−	−	−	−	−	−	−	−	+
Ganoderma sp.	−	−	−	+	−	−	−	+	−	+
Hebeloma cylindrosporum	−	−	+	−	−	−	−	−	−	−
Hebeloma leucosarx	−	−	+	−	−	−	−	−	−	−
Hebeloma sp.	−	−	+	+	−	−	−	−	−	−
Helvellosebacina concrescens	−	−	−	−	−	−	−	−	+	+
Helvellosebacina sp.	−	−	−	−	−	−	+	+	−	+
Hymenogaster griseus	−	−	−	−	−	−	−	+	−	+
Hymenogaster rehsteineri	−	−	−	−	−	−	−	−	−	+
Hymenogaster sp.	−	−	−	−	−	−	−	−	−	+
Hypholoma fasciculare	−	−	−	+	−	−	−	−	−	−
Inocybe mixtilis	−	−	−	−	−	−	−	−	−	+
Inocybe sp.	−	+	−	−	−	−	−	−	−	−
Laccaria sp.	−	−	−	−	−	−	−	+	−	−
Lactarius circellatus	−	−	−	−	−	−	−	+	−	−
Lactarius torminosus	−	−	−	−	+	−	−	−	−	−
Lactarius sp.	−	−	−	−	+	−	−	−	−	−
Leccinum pseudoscabrum	−	−	−	−	−	−	+	+	−	−
Lycoperdon pyriforme	−	−	−	+	−	−	−	−	−	−
Lycoperdon sp.	−	−	−	+	−	−	−	−	−	−
Macrolepiota procera	−	−	−	−	−	−	−	+	−	−
Mallocybe sp.	−	+	−	−	−	−	−	−	−	−
Mycena citrinomarginata	−	−	−	−	−	−	+	+	+	−
Mycena leptocephala	−	−	−	−	−	−	+	+	−	−
Mycena olivaceomarginata	−	−	−	−	−	−	+	−	+	−
Mycena plumipes	−	−	−	−	−	−	+	+	−	−
Mycena pura	−	−	−	−	−	+	−	−	−	−
Mycena sanguinolenta	−	−	+	+	−	−	−	−	−	−
Mycena vulgaris	−	−	−	−	+	+	−	−	−	−
Mycena sp.	−	−	+	−	+	+	+	+	+	+
Mycenella bryophila	−	−	−	−	−	−	−	−	+	−
Mycenella trachyspora	−	−	−	−	−	−	−	−	+	−
Mycenella sp.	−	−	−	−	−	−	−	−	+	−
Odontia fibrosa	−	−	−	+	−	−	−	−	−	−
Odontia sp.	−	−	−	+	−	−	−	−	−	−
Phallus impudicus	−	−	−	+	−	−	−	+	−	+
Phallus ultraduplicatus	−	−	−	+	−	−	−	−	−	−
Phallus sp.	−	−	−	+	−	−	−	−	−	−
Piloderma byssinum	+	−	−	+	−	−	+	+	−	−
Psathyrella candolleana	−	−	−	−	−	−	−	−	−	+
Psathyrella sp.	−	−	−	−	−	−	−	−	−	+
Russula persicina	−	−	−	−	−	−	−	−	+	−
Russula solaris	−	−	−	−	−	−	+	+	+	+
Russula versicolor	−	−	−	−	−	−	+	−	−	−
Russula sp.	−	−	−	−	−	−	+	+	+	+
Saitozyma podzolica	−	−	−	+	−	−	+	+	−	+
Saitozyma sp.	−	−	−	−	−	−	−	+	−	+
Scleroderma bovista	−	−	−	−	−	−	−	−	−	+
Scleroderma sp.	−	−	−	−	−	−	−	−	−	+

Table 1. *Cont.*

The Most Abundant Strains of Fungi Detected in the Roots of Orchids and Surrounding Soils	1		2		3		4		5	
	Epipactis atrorubens		*Epipactis atrorubens*		*Platanthera bifolia*		*Epipactis pontica*		*Cephalanthera longifolia*	
	Ni-Co Mining Heap Dobšiná		Pb-Zn Mining Heap Jasenie				Fe Mining Heap Sirk			
	pH 7.73		pH 7.36		pH 7.81		pH 4.98		pH 6.58	
	Roots	Soil	Roots	Soil	Roots	Soil	Roots	Soil	Roots	Soil
Sebacina sp.	−	−	−	−	−	−	+	+	−	−
Solicoccozyma terricola	−	−	−	+	−	−	+	+	−	+
Solicoccozyma sp.	−	−	−	−	−	−	−	+	−	−
Tephrocybe rancida	−	−	−	−	−	−	−	+	−	−
Thanatephorus sp.	+	−	−	−	−	−	−	−	−	−
Thelephora atra	−	−	+	+	−	−	+	+	−	−
Thelephora caryophyllea	−	−	−	+	−	−	−	−	−	−
Tomentella badia	−	−	−	−	−	−	+	+	−	−
Tomentella fuscocinerea	−	−	+	−	−	−	−	−	−	−
Tomentella lilacinogrisea	−	−	−	+	−	−	−	−	−	−
Tomentella pilosa	−	−	−	−	−	−	+	+	−	+
Tomentella sp.	+	+	−	+	−	+	+	+	−	+
Tremella sp.	−	−	−	−	−	−	−	−	−	+
Tricholoma argyraceum	−	−	+	+	−	−	−	−	−	−
Tricholoma sp.	−	−	+	+	−	−	−	−	−	−
Tylospora sp.	−	+	+	−	−	−	−	−	−	−
Chytridiomycota										
Operculomyces laminatus	−	−	−	+	−	−	−	−	−	−
Rhizophlyctis rosea	−	−	−	+	−	−	−	−	−	−
Rhizophydium sp.	−	+	−	−	−	−	−	−	−	−
Spizellomyces pseudodichotomus	−	+	−	−	−	−	−	−	−	−
Monoblepharomycota										
Sanchytrium sp.	−	−	−	+	−	−	−	−	−	+
Mortierellomycota										
Mortierella alpina	−	+	−	−	−	−	−	−	−	+
Mortierella amoeboidea	−	−	−	−	−	−	−	−	+	+
Mortierella beljakovae	−	−	−	−	−	−	−	−	−	+
Mortierella clonocystis	−	−	−	+	−	−	−	+	−	−
Mortierella gamsii	−	−	−	−	−	−	+	+	−	+
Mortierella globulifera	−	−	−	−	−	−	−	+	−	−
Mortierella humilis	−	−	−	+	−	+	−	+	−	−
Mortierella minutissima	−	−	−	+	−	−	+	+	−	−
Mortierella paraensis	−	−	−	−	−	−	−	+	−	−
Mortierella pseudozygospora	−	−	−	−	−	−	−	+	−	+
Mortierella sarnyensis	−	−	−	−	−	−	−	+	−	+
Mortierella zonata	−	−	−	−	−	−	−	+	−	+
Mortierella sp.	−	+	+	+	−	+	+	+	+	+
Mucoromycota										
Absidia cylindrospora	−	−	−	−	−	−	−	−	−	+
Absidia sp.	−	−	−	−	−	−	−	+	−	−
Mucor hiemalis	−	−	−	−	−	−	−	+	−	−
Mucor sp.	−	−	−	+	−	−	−	−	−	−
Umbelopsis isabellina	−	−	−	−	−	+	−	−	−	−
Umbelopsis sp.	−	−	−	+	−	+	−	−	−	−
Zygorhynchus sp.	−	−	−	+	−	−	−	−	−	−
Σ 171 species	6	29	21	62	9	25	36	57	21	44
Faith's phylogenetic diversity	509	2828	989	3378	429	1141	1053	2368	906	2190

The pH value is a key factor affecting the plant, fungal and bacterial communities on the mining dumps. Soil fungi can grow over a broad range of pH [47], however, soil bacteria are more sensitive to low pH [48], so fungi can dominate this pH spectrum. Values of pH varied from acidic (4.98) to slightly alkaline (7.81).

A high number (57) of fungal species, found in the surrounding soil, tolerate acidic pH of 4.98. Regarding roots, it was 37. Similar to our study, fungi like *Cladophialophora* sp., *Cladosporium* sp., *Exophiala* sp., *Mortierella alpina*, *Mortierella* sp., *Mucor* sp., *Mucor hiemalis*, *Penicillium* sp., *Umbelopsis* sp. provided a great tolerance to a higher concentration of some heavy metals in high acidic to neutral pH of the soils [49]. In regard to to pH, we did not observe anything significant or different in comparison to other studies [50–52]. Organisms from extreme habitats have developed strategies for survival and growth in harsh conditions in the production of organic molecules [53]. They have the

presence in the substrate. Therefore, the studied orchids can grow in technogenic or anthropogenic soils and substrates as well as other extreme habitats. The species composition of mycorrhizal fungi in the roots of *E. atrorubens* is very variable in each locality apart from the *Cortinarius saturatus* species, which is present in all examined roots of orchids. No other species had a significant quantitative representation in both localities of this orchid. Our research has shown that *E. atrorubens* can use either one dominant fungus (Dobšiná deposit, up to almost 65% representation of the species *Trichophaea pseudogregaria*) or even a larger number of species of quantitatively under-represented fungi (Jasenie deposit) for its enormous occurrence in these localities.

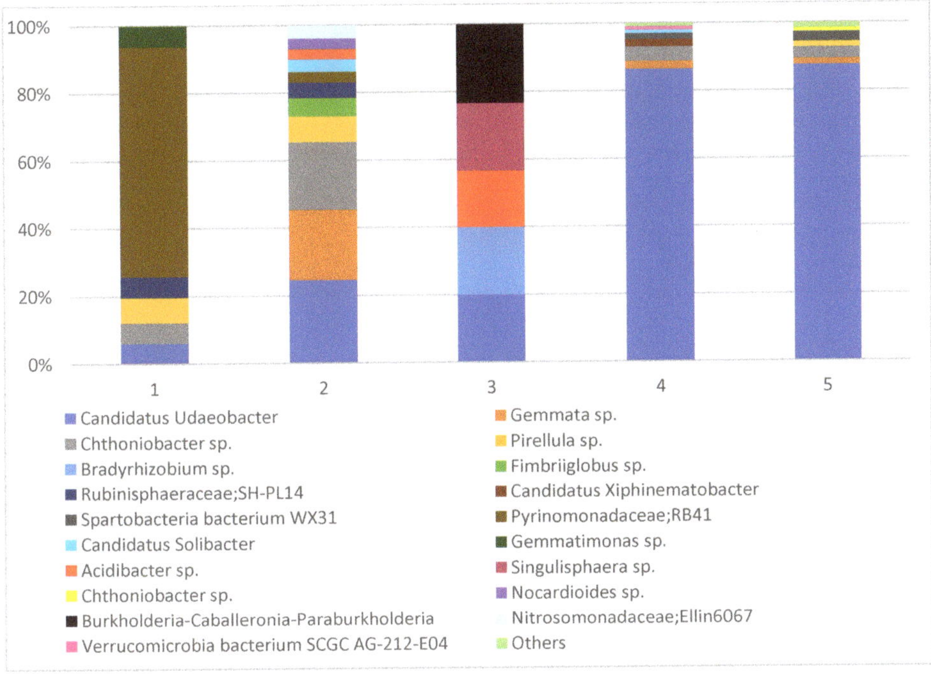

Figure 5. Bacterial communities detected in the surrounding soil of orchids. (1) *E. atrorubens* (Dobšiná); (2) *E. atrorubens* (Jasenie); (3) *P. bifolia* (Jasenie); (4) *E. pontica* (Sirk); (5) *C. longifolia* (Sirk).

4. Conclusions

Due to its adaptation and survival strategy, the family *Orchidaceae* has become an interesting target for microbiome identification. In this work, we identified bacterial and fungal communities from the roots and surrounding soil of orchids from five heavy metal contaminated areas of the mining sites Fe deposit Sirk (*C. longifolia* and *E. pontica*), Pb-Zn deposit Jasenie (*E. atrorubens* and *P. bifolia*), and Ni-Co deposit Dobšiná (*E. atrorubens*). To sum up, slight deviations in pH of the initial soils do not significantly affect the presence of fungi and bacteria and thus the presence of the studied orchids in these localities. Similarly, the toxic elements (e.g., As, Pb, Cr, Ni, Co, Cu, Fe), in the studied localities, do not affect the occurrence of fungi and bacteria, and also orchids. Furthermore, the study of selected factors influencing the occurrence of orchids showed that the most important factors of the occurrence of orchids in technogenic soils are the presence of a sufficient number of fungi and bacteria, sufficient light and the absence of other plant species in the habitat. Moreover, microbial communities also provide a huge benefit for orchids to be able to grow in these polluted areas. Additionally, some of these microbial communities possess huge biotechnological potential in bioremediation of heavy metal contaminated areas. However, to fully understand the microbiota and its interaction with orchids in heavy metal contaminated areas, further investigation is needed.

Author Contributions: Conceptualization, D.O.; methodology, D.O. and T.S.; software, J.B.; validation, M.B.; formal analysis, M.L. and M.B.; investigation, D.O. and M.B.; resources, D.O. and T.S.; data curation, J.B. and M.B.; writing—original draft preparation, M.R. and M.B.; writing—review and editing, D.O. and T.S.; visualization, M.B.; supervision, D.O. and T.S.; project administration, D.O.; funding acquisition, D.O. and T.S. All authors have read and agreed to the published version of the manuscript.

Funding: This research was funded by the Slovak Research and Development Agency (APVV−0375−12 and APVV−16−0264).

Conflicts of Interest: The authors declare no conflict of interest.

References

1. Taylor, D.L.; Bruns, T.D. Independent, specialized invasion of ectomycorrhizal mutualism by two nonphotosynthetic orchids. *Proc. Natl. Acad. Sci. USA* **1997**, *94*, 4510–4515. [CrossRef] [PubMed]
2. McCormick, M.K.; Whigham, D.F.; O'Neill, J.P. Mycorrhizal diversity in photosynthetic terrestrial orchids. *New Phytol.* **2004**, *163*, 425–438. [CrossRef]
3. Gill, D.; Otte, D.; Endler, J.A. Fruiting failure, pollinator inefficiency and speciation in orchids, Speciation and Its Consequences. In *Speciation and Its Consequences*; Otte, D., Endler, J.A., Eds.; Sinauer Associates: Sunderland, MA, USA, 1989; pp. 458–481.
4. Otero, J.T.; Flanagan, N.S. Orchid diversity—Beyond deception. *Trends Ecol. Evol.* **2006**, *21*, 64–65. [PubMed]
5. Banásová, V. Vegetation of copper and antimony mine heaps. *Biol. Práce* **1976**, *22*, 1–109.
6. Rasmussen, H.N. *Terrestrial Orchids: From Seed to Mycotrophic Plant*; Cambridge University Press: Cambridge, UK, 1995; p. 433.
7. George, E.; Häussler, K.U.; Vetterlein, D.; Gorgus, E.; Marschner, H. Water and nutrient translocation by hyphae of *Glomus mosseae*. *Can. J. Bot.* **1992**, *70*, 2130–2137. [CrossRef]
8. Jasper, D.A.; Abbott, L.K.; Robson, A.D. Hyphae of a vesicular-arbuscular mycorrhizal fungus maintain infectivity in dry soil, except when the soil is disturbed. *New Phytol.* **1989**, *112*, 101–107. [CrossRef]
9. Thomas, R.S.; Franson, R.L.; Bethlenfalvay, G.J. Separation of arbuscular mycorrhizal fungus and root effect on soil aggregation. *Soil Sci. Soc. Am. J.* **1993**, *57*, 77–81. [CrossRef]
10. Ashida, J. Adaptation of fungi to metal toxicants. *Annu. Rev. Phytopathol.* **1965**, *3*, 153–174. [CrossRef]
11. Fourest, E.; Roux, J.C. Heavy Metal Biosorption by Fungal Mycelial By-Products Mechanisms and Influence of pH. *Appl. Microbiol. Biotechnol.* **1992**, *37*, 399–403. [CrossRef]
12. Hamilton, W.A. Microbially influenced corrosion as a model system for the study of metal microbe interactions: A unifying electron transfer hypothesis. *Biofouling* **2003**, *19*, 65–76. [CrossRef]
13. Remenár, M.; Harichová, J.; Zámocký, M.; Pangallo, D.; Szemes, T.; Budiš, J.; Šoltys, K.; Ferianc, P. Metagenomics of a nickel-resistant bacterial community in an anthropogenic nickel-contaminated soil in southwest Slovakia. *Biologia* **2017**, *72*, 971–981. [CrossRef]
14. Caporaso, J.G.; Lauber, C.L.; Walters, W.A.; Berg-Lyons, D.; Huntley, J.; Fierer, N.; Owens, S.M.; Betley, J.; Fraser, L.; Bauer, M.; et al. Ultra-high-throughput microbial community analysis on the Illumina HiSeq and MiSeq platforms. *ISME J.* **2012**, *6*, 1621–1624. [CrossRef]
15. Fadrosh, D.W.; Ma, B.; Gajer, P.; Sengamalay, N.; Ott, S.; Brotman, R.M.; Ravel, J. An improved dual-indexing approach for multiplexed 16S rRNA gene sequencing on the Illumina MiSeq platform. *Microbiome* **2014**, *2*, 1–7. [CrossRef] [PubMed]
16. Kraková, L.; Šoltys, K.; Otlewska, A.; Pietrzak, K.; Purkrtová, S.; Savická, D.; Puškárová, A.; Bučková, M.; Szemes, T.; Budiš, J.; et al. Comparison of methods for identification of microbial communities in book collections: Culture-dependent (sequencing and MALDI-TOF MS) and culture-independent (Illumina MiSeq). *Int. Biodeterior. Biodegrad.* **2018**, *131*, 51–59. [CrossRef]
17. Böhmer, M.; Smol'ak, D.; Ženišová, K.; Čaplová, Z.; Pangallo, D.; Puškárová, A.; Bučková, M.; Cabicarová, T.; Budiš, J.; Šoltýs, K.; et al. Comparison of the Microbial Diversity During Two Different Wine Fermentation Processes. *Appl. Sci.* **2020**, *367*, fnaa150. [CrossRef]

18. Hong, C.; Si, Y.; Xing, Y.; Li, Y. Illumina MiSeq sequencing investigation on the contrasting soil bacterial community structures in different iron mining areas. *Environ. Sci. Pollut. Res.* **2015**, *22*, 10788–10799. [CrossRef]

19. Xiao, E.Z.; Krumins, V.; Tang, S.; Xiao, T.F.; Ning, Z.P.; Lan, X.L.; Sun, W.M. Correlating microbial community profiles with geochemical conditions in a watershed heavily contaminated by an antimony tailing pond. *Environ. Pollut.* **2016**, *215*, 141–153. [CrossRef]

20. Račko, M.; Ozdín, D.; Kučerová, G.; Jurkovič, L.; Vaculík, M. Occurrence and uptake of heavy metals by selected terrestrial orchids in extreme conditions of initial soils on previous mining sites. *Biologia* **2020**, in press. [CrossRef]

21. Ďuďa, R.; Ozdín, D. *Minerály Slovenska*; Granit: Praha, Czech Republic, 2012; 480p.

22. Števko, M.; Ozdín, D. Supergene native silver and acanthite from the Jasenie-Soviansko base metals deposit, Nízke Tatry Mts. (Slovak Republic). *Bull. Mineral. Petrolog. Odd. Nár. Muz.* **2012**, *20*, 47–51.

23. Grecula, P. (Ed.) *Mineral Deposits of the Slovak Ore Mountains*; Geocomplex: Bratislava, Slovakia, 1995; Volume 1, 834p.

24. Cao, L.; Qiu, Z.; Dai, X.; Tan, H.; Lin, Y.; Zhou, S. Isolation of Endophytic Actinomycetes From Roots and Leaves of Banana (*Musa Acuminata*) Plants and Their Activities Against *Fusarium oxysporumf*. sp. cubense. *World J. Microbiol. Biotechnol.* **2004**, *20*, 501–504. [CrossRef]

25. Klindworth, A.; Pruesse, E.; Schweer, T.; Peplies, J.; Quast, C.; Horn, M.; Glöckner, F.O. Evaluation of general 16S ribosomal RNA gene PCR primers for classical and next-generation sequencing-based diversity studies. *Nucleic Acids Res.* **2013**, *41*, e1. [CrossRef] [PubMed]

26. White, T.J.; Bruns, T.; Lee, S.; Taylor, J. Amplification and direct sequencing of fungal ribosomal RNA genes for phylogenetics. In *PCR Protocols: A Guide to Methods and Applications*; Innis, M.A., Gelfand, D.H., Sninsky, J.J., White, T.J., Eds.; Academic Press: New York, NY, USA, 1990; pp. 315–321.

27. Bolger, A.M.; Lohse, M.; Usadel, B. Trimmomatic: A flexible trimmer for Illumina sequence data. *Bioinformatics* **2014**, *30*, 2114–2120. [CrossRef] [PubMed]

28. Andrews, S. FastQC: A Quality Control Tool for High Throughput Sequence Data. 2010. Available online: http://www.bioinformatics.babraham.ac.uk/projects/fastqc (accessed on 20 November 2015).

29. Zhang, J.; Kobert, K.; Flouri, T.; Stamatakis, A. PEAR: A fast and accurate Illumina Paired-End reAd mergeR. *Bioinformatics* **2014**, *30*, 614–620. [CrossRef]

30. Bolyen, E.; Rideout, J.R.; Dillon, M.R.; Bokulich, N.A.; Abnet, C.C.; Al-Ghalith, G.A.; Alexander, H.; Alm, E.J.; Arumugam, M.; Asnicar, F.; et al. Reproducible, interactive, scalable and extensible microbiome data science using QIIME 2. *Nat. Biotechnol.* **2019**, *37*, 852–857. [CrossRef]

31. Rognes, T.; Flouri, T.; Nichols, B.; Quince, C.; Mahé, F. VSEARCH: A versatile open source tool for metagenomics. *PeerJ* **2016**, *4*, e2584. [CrossRef] [PubMed]

32. Pedregosa, F.; Varoquaux, G.; Gramfort, A.; Michel, V.; Thirion, B.; Grisel, O.; Blondel, M.; Prettenhofer, P.; Weiss, R.; Dubourg, V.; et al. Scikit-learn: Machine Learning in Python. *J. Mach. Learn. Res.* **2012**, *12*, 2825–2830.

33. Quast, C.; Pruesse, E.; Yilmaz, P.; Gerken, J.; Schweer, T.; Yarza, P.; Peplies, J.; Glöckner, F.O. The SILVA ribosomal RNA gene database project: Improved data processing and web-based tools. *Nucleic Acids Res.* **2013**, *41*, D590–D596. [CrossRef]

34. Nilsson, R.H.; Larsson, K.H.; Taylor, A.F.S.; Bengtsson-Palme, J.; Jeppesen, T.S.; Schigel, D.; Kennedy, P.; Picard, K.; Glöckner, F.O.; Tedersoo, L.; et al. The UNITE database for molecular identification of fungi: Handling dark taxa and parallel taxonomic classifications. *Nucleic Acids Res.* **2019**, *47*, D259–D264. [CrossRef]

35. Faith, D.P. Conservation evaluation and phylogenetic diversity. *Biol. Conserv.* **1992**, *61*, 1–10. [CrossRef]

36. Budiš, J.; Krampl, W.; Kucharík, M.; Hekel, R.; Lichvár, M.; Smoľak, D.; Böhmer, M.; Baláž, A.; Ďuriš, F.; Gazdarica, J.; et al. SnakeLines: Integrated set of computational pipelines for paired-end sequencing reads. 2019. Available online: https://github.com/jbudis/snakelines/ (accessed on 18 October 2019).

37. Köster, J.; Rahmann, S. Snakemake—S scalable bioinformatics workflow engine. *Bioinformatics* **2012**, *28*, 2520–2522. [CrossRef]

38. Turnau, K.; Gawroński, S.; Ryszka, P.; Zook, D. Mycorrhizal-based phytostabilization of Zn–Pb tailings: Lessons from the Trzebionka mining works (Southern Poland). In *Bio-Geo Interactions in Metal-Contaminated Soils*; Soil Biology; Varma, A., Kothe, E., Eds.; Springer: Berlin, Germany, 2012; Volume 31, pp. 327–348.

39. Punz, W.; Sieghardt, H. The response of roots of herbaceous plant species to heavy metals. *Environ. Exp. Bot.* **1993**, *33*, 85–98. [CrossRef]

40. Rufo, L.; de la Fuente, V. Successional dynamics of the climatophile vegetation of the mining territory of the Río Tinto basin (Huelva, Spain): Soil characteristics and implications for phytoremediation. *Arid Land Res. Manag.* **2010**, *24*, 301–327. [CrossRef]

41. Dearnaley, J.D.W.; Martos, F.; Selosse, M.A. Orchid mycorrhizas: Molecular ecology, physiology, evolution and conservation aspects. In *Fungal Associations*, 2nd ed.; The Mycota (9); Hock, B., Ed.; Springer-Verlag: Berlin, Germany, 2012; pp. 207–230.

42. Dearnaley, J.D.W. Further advances in orchid mycorrhizal research. *Mycorrhiza* **2007**, *17*, 475–486. [CrossRef] [PubMed]

43. Shefferson, R.P.; Kull, T.; Tali, K. Mycorrhizal interactions of orchids colonizing Estonian mine tailings hills. *Am. J. Bot.* **2008**, *95*, 156–164. [CrossRef] [PubMed]

44. Ranjan, R.; Rani, A.; Metwally, A.; McGee, H.S.; Perkins, D.L. Analysis of the microbiome: Advantages of whole genome shotgun versus 16S amplicon sequencing. *Biochem. Biophys. Res. Commun.* **2016**, *469*, 967–977. [CrossRef]

45. Knight, R.; Vrbanac, A.; Taylor, B.C.; Aksenov, A.; Callewaert, C.; Debelius, J.; Gonzalez, A.; Kosciolek, T.; McCall, L.I.; McDonald, D.; et al. Best practices for analysing microbiomes. *Nat. Rev. Microbiol.* **2018**, *16*, 410–422. [CrossRef]

46. Quince, C.; Walker, A.W.; Simpson, J.T.; Loman, N.J.; Segata, N. Shotgun metagenomics, from sampling to analysis. *Nat. Biotechnol.* **2017**, *35*, 833–844. [CrossRef]

47. Blagodatskaya, E.V.; Anderson, T.H. Interactive effects of pH and substrate quality on the fungal-to-bacterial ratio and QCO_2 of microbial communities in forest soils. *Soil Biol. Biochem.* **1998**, *30*, 1269–1274. [CrossRef]

48. Fierer, N.; Jackson, R.B. The diversity and biogeography of soil bacterial communities. *Proc. Natl. Acad. Sci. USA* **2006**, *103*, 626–631. [CrossRef]

49. Stępniewska, H.; Uzarowicz, Ł.; Błońska, E.; Kwasowski, W.; Słodczyk, Z.; Gałka, D.; Hebda, A. Fungal abundance and diversity as influenced by properties of Technosols developed from mine wastes containing iron sulphides: A case study from abandoned iron sulphide and uranium mine in Rudki, south-central Poland. *Appl. Soil Ecol.* **2020**, *145*, 103349. [CrossRef]

50. Šimonovičová, A.; Kraková, L.; Pauditšová, E.; Pangallo, D. Occurrence and diversity of cultivable autochthonous microscopic fungi in substrates of old environmental loads from mining activities in Slovakia. *Ecotoxicol. Environ. Saf.* **2019**, *172*, 194–202. [CrossRef] [PubMed]

51. Jurkiewicz, A.; Turnau, K.; Mesjasz-Przybyłowicz, J.; Przybyłowicz, W.; Godzik, B. Heavy metal localisation in mycorrhizas of *Epipactis atrorubens* (Hoffm.) Besser (Orchidaceae) from zinc mine tailings. *Protoplasma* **2001**, *218*, 117–124. [CrossRef] [PubMed]

52. Šimonovičová, A.; Ferianc, P.; Vojtková, H.; Pangallo, D.; Hanajík, P.; Kraková, L.; Feketeová, Z.; Čerňanský, S.; Okenicová, L.; Žemberyová, M.; et al. Alkaline Technosol contaminated by former mining activity and its culturable autochthonous microbiota. *Chemosphere* **2017**, *171*, 89–96. [CrossRef]

53. Timling, I.; Taylor, D.L. Peeking through a frosty window: Molecular insights into the ecology of Arctic soil fungi. *Fungal Ecol.* **2012**, *5*, 419–429. [CrossRef]

54. Chávez, R.; Fierro, F.; García-Rico, R.O.; Vaca, I. Filamentous fungi from extreme environments as a promising source of novel bioactive secondary metabolites. *Front. Microbiol.* **2015**, *6*, 903. [CrossRef]

55. Gupta, H.; Kumar, R.; Park, H.S.; Jeon, B.H. Photocatalytic efficiency of iron oxide nanoparticles for the degradation of priority pollutant anthracene. *Geosyst. Eng.* **2017**, *20*, 21–27. [CrossRef]

56. Mohammadi, K.; Khalesro, S.; Sohrabi, Y.; Heidari, G. A review: Beneficial effects of the mycorrhizal fungi for plant growth. *J. Appl. Environ. Biol. Sci.* **2011**, *1*, 310–319.

57. Baldrian, P. Interactions of heavy metals with white-rot fungi. *Enzym. Microb. Technol.* **2003**, *32*, 78–91. [CrossRef]

58. Zhang, L.; Chen, J.; Gao, C.; Guo, S. *Mycena* sp., a mycorrhizal fungus of the orchid *Dendrobium officinale*. *Mycol. Prog.* **2012**, *11*, 395–401. [CrossRef]

59. Cui, Z.; Zhang, X.; Yang, H.; Sun, L. Bioremediation of heavy metal pollution utilizing composite microbial agent of *Mucor circinelloides*, *Actinomucor* sp. and *Mortierella* sp. *J. Environ. Chem. Eng.* **2017**, *5*, 3616–3621. [CrossRef]

60. Niu, H.; Xu, X.S.; Wang, J.H.; Volesky, B. Removal of lead from aqueous solutions by Penicillium biomass. *Biotechnol. Bioeng.* **1993**, *42*, 785–787. [CrossRef] [PubMed]

61. Chai, B.; Wu, Y.; Liu, P.; Liu, B.; Gao, M. Isolation and phosphate-solubilizing ability of a fungus, *Penicillium* sp. from soil of an alum mine. *J. Basic Microbiol.* **2011**, *51*, 5–14. [CrossRef]

62. Mishra, S.; Chaudhury, G. Biosorption of Copper by *Penicillium* sp. *Int. J.* **1995**, *14*, 111–126. [CrossRef]

63. Gadd, G.M. Metals, Minerals and Microbes: Geomicrobiology and Bioremediation. *Microbiology* **2010**, *156*, 609–643. [CrossRef]

64. Feng, G.; Xie, T.; Wang, X.; Bai, J.; Tang, L.; Zhao, H.; Wei, W.; Wang, M.; Zhao, Y. Metagenomic analysis of microbial community and function involved in cd-contaminated soil. *BMC Microbiol.* **2018**, *18*, 11. [CrossRef]

65. Guo, H.; Nasir, M.; Lv, J.; Dai, Y.; Gao, J. Understanding the variation of microbial community in heavy metals contaminated soil using high throughput sequencing. *Ecotoxicol. Environ. Saf.* **2017**, *144*, 300–306. [CrossRef]

66. Marcin, C.; Marcin, G.; Justyna, M.P.; Katarzyna, K.; Maria, N. Diversity of microorganisms from forest soils differently polluted with heavy metals. *Appl. Soil Ecol.* **2013**, *64*, 7–14.

67. Tipayno, S.C.; Truu, J.; Samaddar, S.; Truu, M.; Preem, J.K.; Oopkaup, K.; Espenberg, M.; Chatterjee, P.; Kang, Y.; Kim, K.; et al. The bacterial community structure and functional profile in the heavy metal contaminated paddy soils, surrounding a nonferrous smelter in South Korea. *Ecol. Evol.* **2018**, *8*, 6157–6168. [CrossRef]

68. Ma, Y.; Wang, Y.; Chen, Q.; Li, Y.; Guo, D.; Nie, X.; Peng, X. Assessment of heavy metal pollution and the effect on bacterial community in acidic and neutral soils. *Ecol. Indic.* **2020**, *117*, 106626. [CrossRef]

69. Brewer, T.E.; Handley, K.M.; Carini, P.; Gilbert, J.A.; Fierer, N. Genome reduction in an abundant and ubiquitous soil bacterium '*Candidatus Udaeobacter copiosus*'. *Nat. Microbiol.* **2016**, *2*, 16198. [CrossRef]

70. Glöckner, F.O.; Kube, M.; Bauer, M.; Teeling, H.; Lombardot, T.; Ludwig, W.; Gade, D.; Beck, A.; Borzym, K.; Heitmann, K.; et al. Complete genome sequence of the marine planctomycete *Pirellula* sp. strain 1. *Proc. Natl. Acad. Sci. USA* **2003**, *100*, 8298–8303. [CrossRef] [PubMed]

71. Turner, T.R.; James, E.K.; Poole, P.S. The plant microbiome. *Genome Biol.* **2013**, *14*, 1267–1273. [CrossRef] [PubMed]

72. Yasouri, F.N. Plasmid mediated degradation of diazinon by three bacterial strains *Pseudomonas* sp., *Flavobacterium* sp. and *Agrobacterium* sp. *Asian J. Chem.* **2006**, *18*, 2437–2444.

73. Lakshmi, C.V.; Kumar, M.; Khanna, S. Biotransformation of chlorpyrifos and bioremediation of contaminated soil. *Int. Biodeter. Biodegr.* **2008**, *62*, 204–209. [CrossRef]

74. Sun, W.; Dong, Y.; Gao, P.; Fu, M.; Ta, K.; Li, J. Microbial communities inhabiting oil-contaminated soils from two major oilfields in Northern China: Implications for active petroleum-degrading capacity. *J. Microbiol.* **2015**, *53*, 371–378. [CrossRef]

75. Chaudhary, D.K.; Kim, J. *Sphingomonas naphthae* sp. nov., isolated from oil-contaminated soil. *Int. J. Syst. Evol. Microbiol.* **2016**, *66*, 4621–4627. [CrossRef]

76. Chen, C.; Liu, Q.; Liu, C.; Yu, J. Effect of different enrichment strategies on microbial community structure in petroleum-contaminated marine sediment in Dalian, China. *Mar. Pollut. Bull.* **2017**, *117*, 274–282. [CrossRef] [PubMed]

77. Zhao, L.; Zhang, C.; Li, H.; Bao, M.; Sun, P. Regulation of different electron acceptors on petroleum hydrocarbon biotransformation to final products in activated sludge biosystems. *Bioprocess Biosyst. Eng.* **2019**, *42*, 643–655. [CrossRef]

78. Feng, X.; Liu, Z.; Jia, X.; Lu, W. Distribution of Bacterial Communities in Petroleum-Contaminated Soils from the Dagang Oilfield, China. *Trans. Tianjin Univ.* **2020**, *26*, 22–32. [CrossRef]

79. Elshahed, M.S.; Youssef, N.H.; Luo, Q.; Najar, F.Z.; Roe, B.A.; Sisk, T.M.; Bühring, S.I.; Hinrichs, K.U.; Krumholz, L.R. Phylogenetic and Metabolic Diversity of Planctomycetes from Anaerobic, Sulfide- and Sulfur-Rich Zodletone Spring, Oklahoma. *Appl. Environ. Microbiol.* **2007**, *73*, 4707–4716. [CrossRef] [PubMed]

80. Arp, D.; Sayavedra-Soto, L.; Hommes, N. Molecular biology and biochemistry of ammonia oxidation by *Nitrosomonas europaea*. *Arch. Microbiol.* **2002**, *178*, 250–255. [CrossRef] [PubMed]
81. Coombs, J.T.; Franco, C.M.M. Isolation and identification of actinobacteria isolated from surface-sterilized wheat roots. *Appl. Environ. Microbiol.* **2003**, *69*, 5603–5608. [CrossRef] [PubMed]
82. Chaudhary, D.K.; Kim, J. *Flavobacterium olei* sp. nov., a novel psychrotolerant bacterium isolated from oil-contaminated soil Free. *Int. J. Syst. Evol. Microbiol.* **2017**, *67*, 2211–2218. [CrossRef] [PubMed]

Publisher's Note: MDPI stays neutral with regard to jurisdictional claims in published maps and institutional affiliations.

Article

Effect of Arbuscular Mycorrhizal Fungi (AMF) and Plant Growth-Promoting Bacteria (PGPR) Inoculations on *Elaeagnus aangustifolia* L. in Saline Soil

Jing Pan [1,2], Cuihua Huang [1], Fei Peng [1,3], Wenjuan Zhang [1,2], Jun Luo [4], Shaoxiu Ma [1] and Xian Xue [1,*]

1   Drylands Salinization Research Station, Key Laboratory of Desert and Desertification, Northwest Institute of Eco-Environment and Resources, Chinese Academy of Sciences, Lanzhou 730000, China; panjing@lzb.ac.cn (J.P.); hch@lzb.ac.cn (C.H.); pengfei@lzb.ac.cn (F.P.); zhangwenjuan@lzb.ac.cn (W.Z.); shaoxiuma@gmail.com (S.M.)
2   University of Chinese Academy of Sciences, Beijing 100049, China
3   International platform for dryland research and education, arid land research center, Tottori University, Tottori 680-8550, Japan
4   School of land and resources, China west normal university, Nanchong 637002, China; luojunxmx@126.com
*   Correspondence: xianxue@lzb.ac.cn; Tel.: +86-931-4967567

Received: 30 December 2019; Accepted: 28 January 2020; Published: 1 February 2020

Abstract: Arbuscular mycorrhizal fungi (AMF) and plant growth-promoting rhizobacteria (PGPR) are considered highly-efficient agents for conferring salt tolerance in host plants and improving soil fertility in rhizosphere. However, information about the inoculation of beneficial microbes on halophytes in arid and semi-arid regions remains inadequate. The objective of this study was to evaluate the influence of AMF (*Glomus mosseae*) inoculation, alone or in combination with PGPR (*Bacillus amyloliquefaciens*), on biomass accumulation, morphological characteristics, photosynthetic capacity, and rhizospheric soil enzyme activities of *Elaeagnus angustifolia* L., a typical halophyte in the northwest of China. The results indicate that, for one-year-old seedlings of *Elaeagnus angustifolia* L., AMF significantly promoted biomass accumulation in aboveground organs, increased the numbers of leaves and branches, and improved the leaf areas, stem diameters and plant height. AMF-mediated morphological characteristics of aboveground organs favored light interception and absorption and maximized the capacities for photosynthesis, transpiration, carbon dioxide assimilation and gas exchange of *Elaeagnus angustifolia* L. seedlings in saline soil. AMF also promoted root growth, modified root architecture, and enhanced soil enzyme activities. *Elaeagnus angustifolia* L. was more responsive to specific inoculation by AMF than by a combination of AMF and PGPR or by solely PGPR in saline soils. Therefore, we suggest that *G. mosseae* can be used in saline soil to enhance *Elaeagnus angustifolia* L. seedlings growth and improve soil nutrient uptake. This represents a biological technique to aid in restoration of saline-degraded areas.

Keywords: halophyte; arbuscular mycorrhizal fungi; plant growth promoting rhizobacteria; morphological characteristics; photosynthesis; soil enzymes

1. Introduction

Soil salinity is a severe agronomical, ecological, and socioeconomic concern in most arid and semi-arid regions of the world [1,2]. Most of salt-affected land has become saline owing to natural accumulation of salts over long periods of time, while others are caused by anthropogenic mismanagement [1]. Salinity in soil provokes the loss of natural plant communities [2], destroys ecosystem diversity [3], reduces the production of crops and forages [4], accelerates the process of soil degradation [2], and reduces the potential utilization of halophytes as crops [5]. Salinity not only adversely affects agriculture and ecosystems, but also poses a threat to public health and security [6], which further exacerbates unsustainable livelihoods and inequalities in socio-economic development [7]. Moreover, saline soil is sensitive to global climate change, this further intensifies salinization [8,9]. It is estimated that salinization will threaten more than 50% of arable land worldwide by 2050 [10]. Hence, this silent hazard will continue to threaten agricultural sustainability, food security, ecosystem stability, human health and income generation.

The northwest arid area of China is one such region that is highly vulnerable to global climate change, and is widely affected by soil salinization [11]. The Minqin Oasis is considered an essential part of the Chinese Silk Road in the northwest arid area of China. As one of the world's driest and most fragile regions, the Minqin oasis has witnessed significant impacts of both natural and anthropogenic factors on soil salinization. Harsh natural conditions, such as very high temperatures, low rainfall, and strong evaporation, lead to soil natural salinization [11,12]. Many organism inhabitants are well well adapted to this naturally saline environment. Anthropogenic factors, such as outdated irrigation systems combined with excessive groundwater extraction, execrate soil secondary salinization. Secondary salinization inhibits plant growth, reduces crop yield, changes the habitat pattern of halophytes, and causes vegetation degeneration and ecosystem deterioration in the Minqin Oasis [13–15], which further results in wasted land. If this wasted land is not improved in a timely manner, Minqin oasis will become a new source of salt dust and a national ecological crisis zone in China. Therefore, strengthening saline soil research in the Minqin Oasis not only has a positive effect on the regional biodiversity conservation, but also guides the study of ecosystems in the northwest arid areas of China.

Elaeagnus angustifolia L., a very important halophytic perennial tree in the Minqin Oasis, has excellent tolerance to salinization, sand and barren land [16], and has high value in ecology, economics and medicine [17,18]. However, the persistent expansion and aggravation of soil salinization in the Minqin Oasis, driven by both natural and anthropogenic factors, is threatening the habitats of *Elaeagnus angustifolia* L. Salinity intrusion in this area not only adversely affects the growth, productivity and distribution of *Elaeagnus angustifolia* L. but also lowers the sheltering effects of the protective forest dominated by *Elaeagnus angustifolia* L. to farmland [16]. Previous studies have reported that indigenous halophytes can survive in saline habitats thanks to the co-evolution of halophytes and microbes [2,19,20]. The collaboration of halophytes and microbes not only enhances plant biomass and population diversity [21,22] but also improves soil microbial activities and other soil properties [19,23]. Therefore, there is potential to exploit beneficial soil microbes to promote the growth of *Elaeagnus angustifolia* L. and to further use these microbes to ameliorate saline soil in the Minqin Oasis.

Arbuscular mycorrhizal fungi (AMF) and plant growth-promoting rhizobacteria (PGPR) are beneficial below-ground microbes that directly associate with plant roots [24,25]. Previous studies showed that the growth-promoting effects on host glycophytes under salt stress are attributable to AMF and PGPR, which could enhance salt tolerance by altering root morphology, modifying root-to-shoot communication, increasing nutrients uptake, maintaining ion homeostasis, decreasing oxidative damage, and elevating photosynthetic capacity [10,24,26,27]. As bio-inoculants, AMF and PGPR also play vital roles in soil biological activity, benefiting the remediation of saline soil, and consequently increasing the suitability of saline soil to plant growth [28,29]. However, little is known about the inoculation effects of different beneficial microorganisms on the biomass accumulation, morphological characteristics, photosynthetic capacities, and rhizospheric soil enzymes of halophytes, especially *Elaeagnus angustifolia* L. in saline soil.

Based on this existing knowledge, the objective of our study is to test the effects of AMF inoculation, both alone and in combination with PGPR, on *Elaeagnus angustifolia* L. in saline soil. We address the following research questions: (1) what are the impacts of AMF and PGPR inoculations on biomass accumulation, morphological characteristics, photosynthesis and soil enzymes of *Elaeagnus angustifolia*

L. in saline soils? (2) Is species diversity or a particular key species most important in controlling the growth characteristics and salt tolerance of *Elaeagnus angustifolia* L.?

2. Materials and Methods

2.1. Study Site

The study area is located at the Drylands Salinization Station affiliated to the Northwest Institute of Eco-Environment and Resources, Chinese Academy of Sciences, China, which is located in the low reaches of the Shiyang River (N 39°02′38″, E 103°36′36″) (Figure 1).

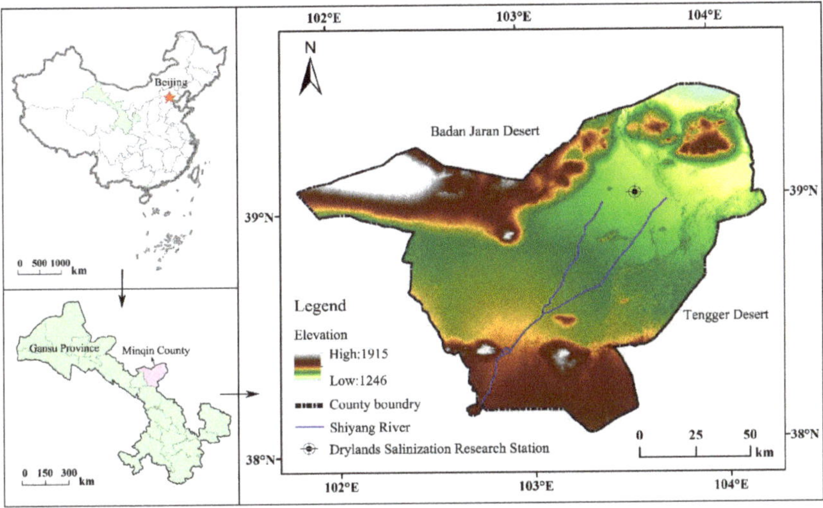

Figure 1. The location of the Drylands Salinization Station.

Over a 30-year period, the mean annual precipitation is 119.46 mm. The maximum annual precipitation is 202 mm in 1994, and the minimum annual precipitation is 77.3 mm in 1989 (Figure 2a). In 2007, the total precipitation is 95 mm. The maximum temperature is 42.13 °C in July, and minimum temperature is −25.9 °C in January. The maximum and minimum photosynthetic active radiation is 730.2 μmol m^{-2}s^{-1} and 41.57 μmol m^{-2}s^{-1}, respectively (Figure 2b).

2.2. Soil and Plant Treatment

Soil and seeds utilized in this experiment were collected from the Drylands Salinization Station affiliated to the Northwest Institute of Eco-Environment and Resources, Chinese Academy of Sciences, China, which is located in the low reaches of the Shiyang River (N 39°02′38″, E 103°36′36″) (Figure 1). The soil was collected from the top layer (0–20 cm) of an agricultural field, then passed through a 5 mm sieve. The soil was classified into Orthic Halosols in the Chinese taxonomic system and was identified as Ustifluvent in the taxonomic system of United States Department of Agriculture [30,31]. The soil used throughout this experiment before planting had the following properties: pH of 8.14±0.02, EC of 2015 ± 15 μS cm^{-1} (water-soil ratio of 5:1), organic carbon content of 9.36 ± 0.18 g kg^{-1}, total N of 0.61 ± 0.01 g kg^{-1}, available P of 16.41 ± 0.81 kg ha^{-1}, and available K of 47.75 mg kg^{-1} [11]. Healthy fruits of *Elaeagnus angustifolia* L. were selected carefully, and peeled to obtain the seeds. These seeds were surface-sterilized with NaOCl (5%) for 10 min and then thoroughly rinsed with deionized water to completely remove all traces of NaOCl.

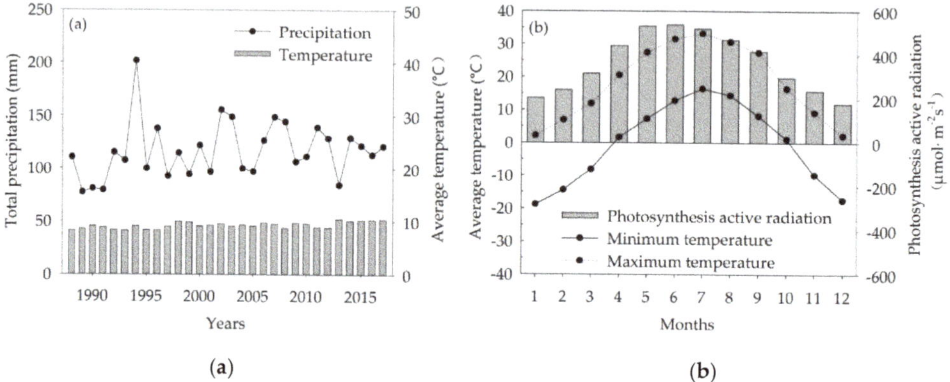

Figure 2. The characteristics of precipitation, temperature and photosynthesis active radiation (from meteorological data in the Drylands Salinization Station). (**a**) total precipitation amount and average temperature per year, (**b**) average temperature and photosynthesis active radiation per month.

2.3. Microbial Inoculum

Glomus mosseae, the most efficient fungus to help plants to cope with the detrimental effects of salt stress [32]. The strain of *G. mosseae* was commercially provided by the Institute of Plant Nutrition and Resources, Beijing Academy of Agriculture and Forestry Sciences, which was isolated from the rhizospheric soil of *Populus euphratica* in Xinjiang Autonomous Region, China. *G. mosseae* was multiplied using *Sorghum bicolor* as host plant and a mixture of soil, river sand, perlite and quartz sand as substrate (with a mixing ratio of 2:3:3:3) for 4 months. The AMF inoculum contained growth substrate, spores (density of approximately 75 per 10 g dry substrate), mycelium and infected root fragments of *G. mosseae*. Each pot was inoculated with 100 g inoculum for mycorrhizal treatment. Non-mycorrhizal treatment received 100 g sterilized AM inoculum (160 °C, 4h). *Bacillus amyloliquefaciens* FZB42 was grown overnight on Luria–Bertani liquid medium with constant shaking at 200 rpm [33]. After incubating overnight, cells were obtained by centrifugation at 10,000 rpm for 6 min, washed once, and re-suspended in deionized water to 1.0×10^8 colony-forming units/mL for use as the PGPR inoculum [33,34].

2.4. Plant Growth Conditions and Experimental Design

A pot experiment was performed outdoors under natural conditions in the Drylands Salinization Station during 11 May to 29 August 2017. Five seeds were sown in experimental pots filled with 1.3 kg saline soil. The pot bottom was covered by mesh (38 μm) to avoid root growth out of the pot. The experiment consisted of four treatments as follows: (1) Control: seeds were soaked in 40 ml sterile water for 30 min and the soil was mixed with sterile AMF inoculum. (2) AM treatment: seeds were soaked in 40 ml sterile water for 30 min and soil was mixed with AMF inoculum. (3) PGPR treatment: seeds were inoculated with PGPR inoculum for 30 min, soil was mixed with sterile AMF inoculum. (4) Co-microbial treatment: soil was treated following the AMF treatment; seeds were treated following the PGPR treatment. After 15 days of emergence, thinning was performed and one seedling was left in each pot. Seedlings had roughly equal heights. Then, seedlings in the PGPR treatment and co-microbial treatment were irrigated with 40 mL PGPR inoculum (10^8 CFU/ml). All plants were regularly irrigated with water from Shiyang River to maintain soil moisture near field capacity. Each treatment contained eight pots and the seedlings were harvested 110 days after sowing. Five pots were used to measure the parameters of plant biomass, organ morphology, and photosynthesis, and the remaining three pots were used to determine soil enzymes.

2.5. Parameter Measurements

2.5.1. Plant Biomass, Organ Morphology, and AMF Colonization Rate

Five replicates of plant height and leaf number were measured and recorded before the end of the experiment. The basal stem diameter was determined by a Vernier caliper. Total leaf area was scanned by an Epson Expression 1480 digital scanner (Epson (China) Co., Beijing, China) and measured by the measurement tool in Adobe Photoshop CS4 software (Version 11.0.1). Plant fresh mass was measured at final harvest, plant dry mass was determined and calculated after oven drying at 60 °C to constant mass [35]. The degree of succulence was expressed as the ratio of fresh mass to dry mass [36,37]. Specific stem length was calculated as the ratio of plant height to stem dry mass [38]. AMF colonization rate was also measured at final harvest using fresh root segments which was cut into 1 cm length [39].

The fresh roots were carefully washed under running tap water, dried with paper towels and scanned using the digital scanner Epson Expression 1480 (Epson (China) Co., Beijing, China) to analyze root total length, surface area, projective area, and tip numbers using the WinRHIZO software (Regent Instruments Inc., Quebec, QC, Canada). Root volume was measured in a graduated cylinder by the drainage method. The specific root length was calculated as root length divided by the root dry mass [40].

2.5.2. Photosynthesis Measurement

Net photosynthetic rates (Pn), stomatal conductance (Gs), transpiration rate (Tr), and carboxylation efficiency (CE) were determined on the fully expanded leaves using a portable open-flow gas exchange system LI-6400XT (LI-COR, Lincoln, NE, U.S.A) during the period from 9:30 to 11:30 am under natural sunlight and CO_2 (400 μmol mol^{-1}) on a sunny day before harvest [41].

2.5.3. Determination of Soil Enzymes

The aerial parts of seedlings were removed after harvesting from the remaining three pots. Visible root materials with adhering soil were removed manually. Remaining soil samples used for analyzing enzymes activities were collected and sieved (2 mm) before analysis. The activities of urease (UE) and sucrase (SC) were determined by indophenol blue colorimetry and 3,5-dinitrosalicylic acid method, respectively [42]. Soil alkaline phosphatase (AKP) was determined by disodium phenylphosphate method [43]. Soil catalase (CAT) was determined by colorimetric method conducted by ultraviolet spectrophometry [44]. All enzymes activities were detected using soil enzyme kits according to the manufacturer's instructions (Comin Biotechnology Co., Ltd, Suzhou, China).

2.6. Statistical Analysis

The experimental data were log-transformed and then analyzed by the analysis of variance, and the comparison among treatment means was conducted using Tukey's test at the 5% level with the statistical software package SPSS 22.0 (SPSS for Windows, Version 22.0, Chicago, IL, USA). Non-transformed data appear in all figures, which were drawn by Sigma Plot version 10.0 (Systat Software Inc., Point Richmond, CA, USA).

2.7. Estimation of Salt Tolerance and Soil Enzyme Activities

Salt tolerance of *Elaeagnus angustifolia* L. seedlings and rhizospheric soil enzyme activities were evaluated by the subordinate function value (SFV) [45] combined with weights. The SFV was calculated using the Equation (1):

$$U\left(X_{ij}\right) = \frac{X_{ij} - X_{j\,min}}{X_{jmax} - X_{j\,min}} \tag{1}$$

where $U(X_{ij})$ is the SFV of the index (j) for the treatment (i), X_{ij} is the actual measured value of the index (j) for the treatment (i), and X_{jmax} and X_{jmin} are the maximum and minimum values of the index (j) observed in the treatment (i).

The standard deviation coefficient method is used to determine the weight of different indexes. The coefficient of standard deviation (V_j) and weight (W_j) were evaluated via the following equations:

$$V_j = \sqrt{\sum_{j=1}^{n}\left(X_{ij} - \overline{X}_j\right)^2 / \overline{X}_j} \tag{2}$$

$$W_j = V_j / \sum_{j=1}^{m} V_j \tag{3}$$

where X_{ij} is the actual measured value of the index (j) for the treatment (i), $\overline{X}_j$ is the mean value of the index (j). Finally, the plant salt tolerance and rhizospheric soil fertility of different treatments are expressed by the comprehensive evaluation value (D). We then used Equation (4) to calculate the comprehensive evaluation value (D):

$$D = \sum_{j=1}^{n}\left(U(X_j) \times W_j\right) \tag{4}$$

where $U(X_j)$ were the SFV of the index (j). D ranges from 0 to 1.

3. Results

3.1. Effects on Biomass Accumulation and Arbuscular Mycorrhizal Fungi (AMF) Colonization Rate

Leaf and stem dry masses in the AMF treatment and co-microbial treatment were markedly higher than those in the control (Figure 3a,b). Microbial inoculations significantly increased root dry mass when compared with the control (Figure 3c). AMF colonization rates ranged from 5% in non-inoculated plants to 89.2% in single AMF-inoculated plants, and there are also significant differences between the three microbial treatments (Figure 3d).

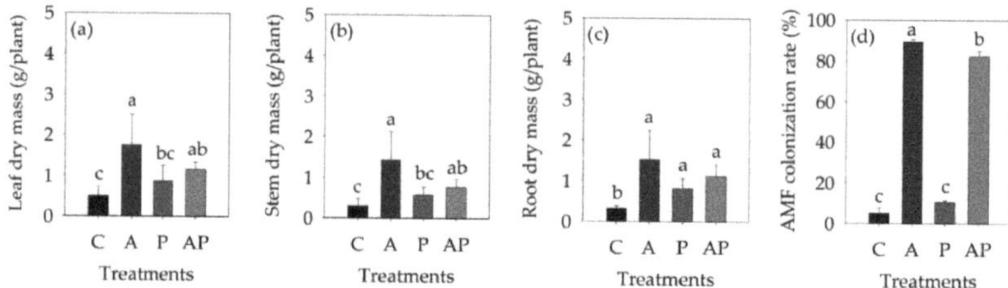

Figure 3. Effects of *Glomus mosseae* and *Bacillus amyloliquefaciens* on biomass and arbuscular mycorrhizal fungi (AMF) colonization rate of *Elaeagnus angustifolia* L. seedlings in saline soil. Values are the means ± standard deviation of five replicates. Value sharing the different letters indicate significant differences between treatments at the 5% level (Tukey's t-test). C: Control, A: AMF (*Glomus mosseae*) treatment, P: plant growth-promoting rhizobacteria (PGPR) (*Bacillus amyloliquefaciens*) treatment, AP: AMF + PGPR treatment. (**a**) leaf dry mass, (**b**) stem dry mass, (**c**) root dry mass, (**d**) AMF colonization rate.

3.2. Effects on Morphological Characteristics of Aboveground Organs

Leaf numbers and total leaf areas in three microbial treatments statistically increased, while specific stem length in microbial treatments significantly decreased in saline soil when compared with the control (Figure 4a,b,f). Only a single AMF inoculation significantly improved plant height when compared with the control, by 64% (Figure 4c). The presence of AMF, whether singly or combined with PGPR, statistically increased basal stem diameters and branch numbers when compared with the corresponding controls (Figure 4d,e). Both leaf succulence and stem succulence significantly decreased in single AMF-treated plants when compared to the control, by 28.04% and 27.72%, respectively (Figure 4g,h).

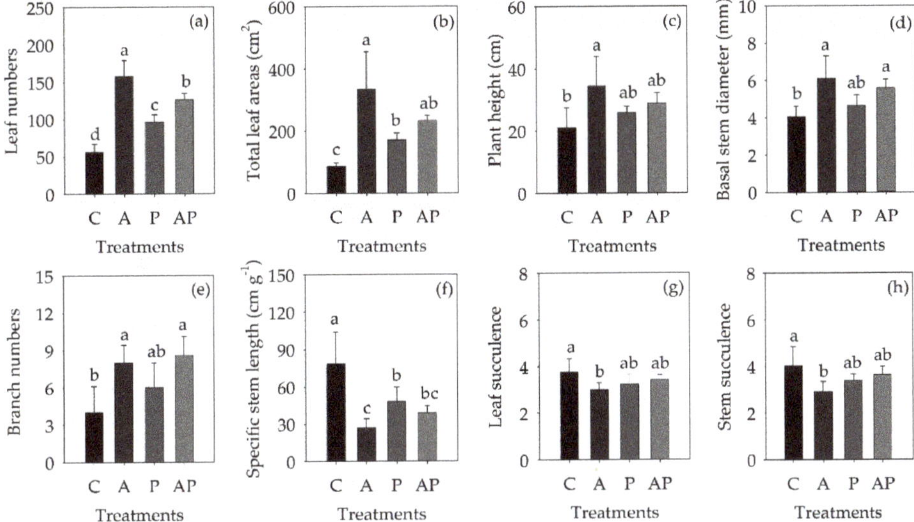

Figure 4. Effects of *Glomus mosseae* and *Bacillus amyloliquefaciens* on aboveground organ morphological characteristics of *Elaeagnus angustifolia* L. seedlings in saline soils. Values are the means ± standard deviation of five replicates. Value sharing the different letters indicate significant differences between treatments at the 5% level (Tukey's *t*-test). C: Control, A: AMF (*Glomus mosseae*) treatment, P: PGPR (*Bacillus amyloliquefaciens*) treatment, AP: AMF + PGPR treatment. (**a**) Leaf numbers, (**b**) total leaf areas, (**c**) plant height, (**d**) basal stem diameter, (**e**) branch numbers, (**f**) specific stem length, (**g**) leaf succulence, (**h**) stem succulence.

3.3. Effects on Morphological Characteristics of Roots

Root length significantly increased in the AMF treatment, by 43.89% when compared with the control, and was also markedly higher than that in the PGPR treatment (Figure 5a). Root surface area and root projected area in the PGPR treatments were considerably lower than those in the AMF treatment and co-microbial treatment (Figure 5b,c). The presence of AMF, singly or combined with PGPR, significantly increased the root average diameters and root tip numbers (Figure 5d,e). Compared to the control, all three microbial inoculations significantly increased root volumes (by 58.9–185.04%) (Figure 5f), but markedly decreased specific root lengths (by 54.74–64.38%) and root succulence (by 33.21–49.31%) (Figure 5g,h).

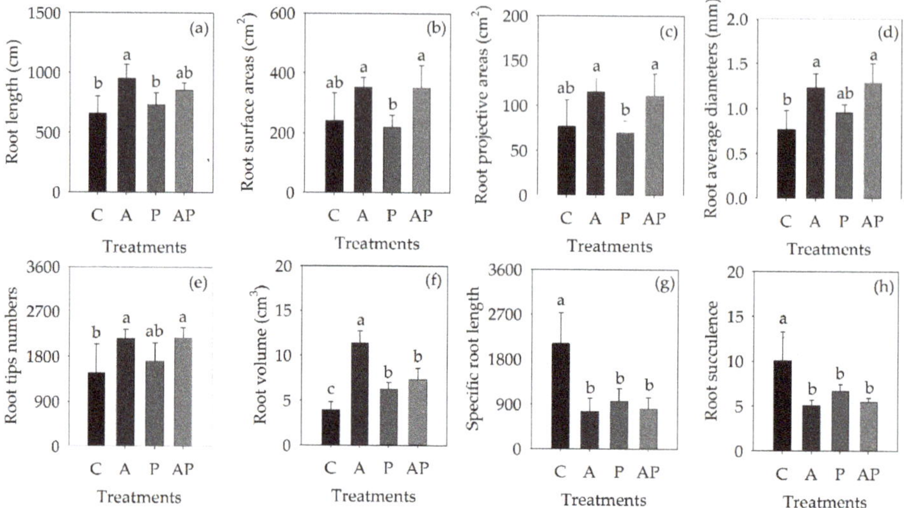

Figure 5. Effects of *Glomus mosseae* and *Bacillus amyloliquefaciens* on root morphological characteristics of *Elaeagnus angustifolia* L. seedlings in saline soils. Values are the means ± standard deviation of five replicates. Value sharing the different letters indicate significant differences between treatments at the 5% level (Tukey's *t*-test). C: Control, A: AMF (*Glomus mosseae*) treatment, P: PGPR (*Bacillus amyloliquefaciens*) treatment, AP: AMF + PGPR treatment. (**a**) root length, (**b**) root surface areas, (**c**) root projective areas, (**d**) root average diameters, (**e**) root tips numbers, (**f**) root volume, (**g**) specific root length, (**h**) root succulence.

3.4. Effects on Photosynthesis

Pn and Gs in three microbial treatments significantly increased when compared with the corresponding controls under saline soil conditions, and significant differences were evident among the three microbial-inoculated treatments (Figure 6a,c). The presence of AMF, whether singly or combined with PGPR, statistically increased Tr (Figure 6b). CE only significantly increased in the AMF treatment when compared with control, by 40.84%. CE in the AMF treatment was also significantly higher than that in the PGPR treatment and co-microbial treatment, by 40.69% and 29.82%, respectively (Figure 6d).

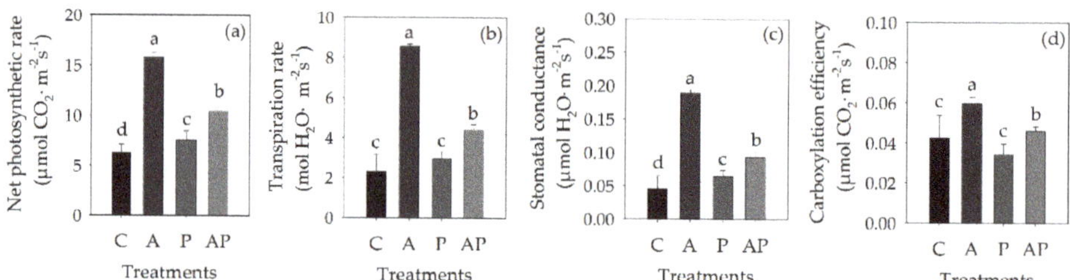

Figure 6. Effects of *Glomus mosseae* and *Bacillus amyloliquefaciens* on photosynthetic parameters of *Elaeagnus angustifolia* L. seedlings in saline soils. Values are the means ± standard deviation of five replicates. Value sharing the different letters indicate significant differences between treatments at the 5% level (Tukey's *t*-test). C: Control, A: AMF (*Glomus mosseae*) treatment, P: PGPR (*Bacillus amyloliquefaciens*) treatment, AP: AMF + PGPR treatment. (**a**) net photosynthetic rate, (**b**) transpiration rate, (**c**) stomatal conductance, (**d**) carboxylation efficiency.

3.5. Effects on Soil Enzymes

The CAT in the rhizosphere soils only significantly increased in the AMF treatment (Figure 7a). AMF inoculation, singly or combined with PGPR, significantly increased AKP in the rhizosphere soil, by an average of 12.86% when compared to the control (Figure 7b). A significant increase of UE was only found in the single AMF inoculation, by 36.01% (Figure 7c). Compared to the control, SC increased by 27.41% and 53.28% in the AMF treatment and co-microbial treatment, respectively (Figure 7d).

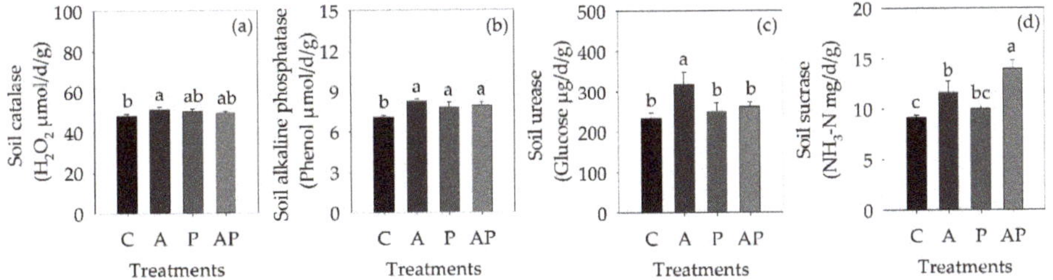

Figure 7. Effects of *Glomus mosseae* and *Bacillus amyloliquefaciens* on photosynthetic parameters of *Elaeagnus angustifolia* L. seedlings in saline soils. Values are the means ± standard deviation of three replicates. Value sharing the different letters indicate significant differences between treatments at the 5% level (Tukey's *t*-test). C: Control, A: AMF (*Glomus mosseae*) treatment, P: PGPR (*Bacillus amyloliquefaciens*) treatment, AP: AMF + PGPR treatment. (**a**) soil catalase, (**b**) soil alkaline phosphatase, (**c**) soil urease, (**d**) soil sucrase.

3.6. Salt Tolerance Assessment

Based on the ranking of SFVs of 18 parameters, the AMF treatment had the highest SFV and control had the lowest. The salt tolerance ranking of different treatments was AMF treatment > AMF and PGPR treatment > PGPR treatment > control (Table 1).

Table 1. Salt tolerance assessment of *Elaeagnus angustifolia* L. under different treatments using the subordinate function value method.

Parameters	C	A	P	AP	W
LDM	0	1	0.2953	0.53	0.0737
SDM	0	1	0.2452	0.4141	0.094
RDM	0	1	0.4084	0.6642	0.0803
BN	0	0.8696	0.4348	1	0.047
BSD	0	1	0.2873	0.7562	0.0269
PH	0	1	0.3661	0.5923	0.0306
LN	0	1	0.3933	0.6917	0.0594
TLA	0	1	0.3384	0.5892	0.0761
RL	0	1	0.2436	0.6783	0.0243
RPA	0.1513	1	0	0.9095	0.0375
RSA	0.1625	1	0	0.9765	0.0363
RAD	0	0.8962	0.3538	1	0.0345
RV	0	1	0.3047	0.4537	0.064
RTN	0	0.9729	0.3531	1	0.0274
Pn	0	1	0.1293	0.4327	0.0633
Gs	0	1	0.1325	0.3356	0.0977
Tr	0	1	0.0974	0.3323	0.0922
CE	0.6827	1	0	0.4585	0.0347
D	0.035262	0.989448	0.243499	0.586412	
Sequence	4	1	3	2	

Note: C: Control, A: AMF (*Glomus mosseae*) treatment, P: PGPR (*Bacillus amyloliquefaciens*) treatment, LDM: leaf dry mass, SDM: stem dry mass, RDM: root dry mass, BN: brunch number, BSD: basal stem diameter, PH: plant height, LN: leaf numbers, TLA: total leaf area, RL: root length, RPA: root projective area, RSA: root surface area, RD: root diameter, RV: root volume, RTN: root tips number, Pn: net photosynthetic capacity, Gs: gas conductance, Tr: transpiration rate, CE: carboxylation efficiency, D: comprehensive evaluation value, W: weight.

3.7. Soil Fertility Assessment

We used the SFV method to evaluate the soil fertility in the rhizosphere of *Elaeagnus angustifolia* L. seedlings in saline soils. Based on the ranking of SFVs, the soil enzymes ranking was AMF treatment > co-microbial treatment > PGPR treatment > control (Table 2).

Table 2. Soil fertility assessment in the rhizosphere of *Elaeagnus angustifolia* L. seedlings under different treatments using the subordinate function value method.

Treatment	C	A	P	AP	W
CAT	0	1	0.6989	0.5223	0.064
AKP	0	1	0.6345	0.7171	0.1514
SC	0	0.5144	0.1817	1	0.3289
UE	0	1	0.1885	0.3396	0.4556
D	0	0.8402	0.2864	0.6256	
Sequence	4	1	3	2	

Note: C: Control, A: AMF (*Glomus mosseae*) treatment, P: PGPR (*Bacillus amyloliquefaciens*) treatment, AP: AMF + PGPR treatment, CAT: Soil catalase, AKP: Soil alkaline phosphatase, SC: Soil sucrase, UE: Soil urease, *D*: comprehensive evaluation value, W: weight.

4. Discussion

Soil salinity induces adverse effects on seedling establishment and biomass accumulation [1,46]. Although halophytes possess inherent salt-tolerant strategies, excessive soil salinity could damage seedling establishment and further inhibit growth in halophytes [47]. Soil microbes are of pivotal importance for plant growth, especially in adverse ecosystems such as those with saline soil conditions [48]. Previous studies reported that AMF and PGPR inoculations could reduce biomass losses induced by soil salinity, and that co-microbial inoculation performed better than single AMF or PGPR inoculations [49–52]. In contrast, the results in this study show that the single AMF inoculation promoted significantly greater biomass accumulation and AMF colonization rate than that in the non-inoculated control, and was even slightly greater than that in the co-microbial inoculation (albeit there was no statistical difference) (Figure 3). Thus, *Elaeagnus angustifolia* L. seedlings are more responsive to the specific AMF than to overall microbial richness or diversity in saline soils. Therefore, key specific microbial species, not the microbial richness or diversity, determined biomass accumulation in our study [53].

The *Bacillus amyloliquefaciens* FZB42 strain was used in this experiment as the PGPR inoculum. Previous studies showed that FZB42 could stimulate fresh and dry mass accumulation and induce systemic salt tolerance of *Arabidopsis thaliana* under 100 mM NaCl hydroponic culture conditions [33]. However, FZB42 was only effective in improving root biomass accumulation and salt tolerance of *Elaeagnus angustifolia* L. seedlings in the present study under saline soil conditions (Figure 3, Table 1). Hence, individual microbe species vary greatly in their effect on different plant species in different growth conditions. Moreover, both positive and negative effects of PGPR in plants have been reported in previous publications [54–56]. Hence, we suggest that PGPR efficiency of different plants may be influenced either by the growth environment or by the existence of "functional specificity" in the combination of plant and microbe species [48].

The growth and development of halophytes under saline conditions are inseparable from their own adaptive mechanisms, in which the adaptions of their morphological characteristics are the most intuitive [57]. The plant has to adjust its physiological functions and morphological characteristics in the leaf, stem and root to adapt to adverse environments as well as to ensure survival [37,58,59]. Root systems are the first "line of defense" to encounter saline soil, and root-system indicators are often used to quantify the acquisition capacity of water and nutrients in plants [60]. Soil salinity directly influenced root growth [57,59]. Previous studies have documented that AMF and PGPR colonization modulated the root-system architecture and reversed salinity effects on agricultural production both quantitatively and qualitatively in salt-sensitive plants such as soybean, citrus, and maize [26,40,59,61,62]. The present investigation showed that AMF inoculation, singly or mixed with PGPR, significantly increased root biomass, length, surface area, average diameter, and tips number of *Elaeagnus angustifolia* L. seedlings when compared to control and PGPR inoculation (Figures 3 and 5). The root volumes in AMF inoculated seedlings were significantly higher than those in PGPR inoculated seedlings (Figure 5f), which indicates that *Elaeagnus angustifolia* L. seedlings are more responsive to AMF inoculation than to PGPR inoculation in saline soils.

Generally, a higher specific root length indicates higher root hydraulic conductivity and proliferation rate [57]. Root succulence is an adaptive mechanism to improve salt tolerance by diluting salt content in plants [63]. Our results confirmed that microbial inoculation significantly decreased specific root length and root succulence in *Elaeagnus angustifolia* L. seedlings when compared to the control (Figure 5g,h). It is possible that AMF hyphae replace the absorption function of roots, and that specific compounds exuded from PGPR and AMF provide more nutrients and exclude toxic ions from the rhizosphere, allowing the host to develop thicker roots and a larger absorption area [59,64,65]. Thus, *Elaeagnus angustifolia* L. seedlings with AMF and PGPR inoculation did not require greater specific root lengths, either as a surrogate for diameter to obtain more resources [66], or as an adaptive mechanism to compensate for loss of root biomass under osmotic stress induced by salinity [57,59]. Furthermore, *Elaeagnus angustifolia* L. seedlings with AMF and PGPR inoculations also had no requirement for a greater root succulence to reduce the damage caused by toxic ions [63].

The optimization of root morphology facilitated the transmission of soil water and nutrients to the stems, branches and leaves, which benefits the growth of aboveground organs [58]. Stems not only transport water and nutrients from the roots but also bear the full weight of aboveground tissues and provide growth spaces for branches and leaves [58]. Salinity stress suppresses stem and branch growth [67]. In this study, we found that stem biomass, plant height, stem diameter, and branch numbers in *Elaeagnus angustifolia* L. were increased markedly by the presence of AMF (Figures 3 and 4), consistent with previous reports on citrus, fenugreek and trifoliate orange [62,67,68]. However, the specific stem length in *Elaeagnus angustifolia* L. decreased significantly in three microbial treatments (Figure 4f). This result may be interpreted as AMF and PGPR reversing the salinity effects on stems and branches, such that *Elaeagnus angustifolia* L. with microbial inoculation did not require a greater specific stem length to ensure the effectiveness of associated functions. Hence, we suggest that increasing specific stem length may be one of the self-adaptive mechanisms used by *Elaeagnus angustifolia* L. without microbial inoculation to counter adverse salt stress conditions.

As well as the growth of stems and branches, we found that leaf biomass, leaf numbers, and leaf total areas in *Elaeagnus angustifolia* L. seedlings also increased with the presence of AMF (Figures 3 and 4). This result indicates that AMF and PGPR could reduce salt damage on leaves and promote the growth and surface expansion of leaves in *Elaeagnus angustifolia* L. seedlings under saline soil conditions, which is also consistent with studies of citrus [69], alfalfa [70], and beans [71] inoculated with AMF under salt stress conditions. Similar to the results of root succulence, microbial inoculation decreased the succulence of leaf and stem (albeit the differences were significant only in the single AMF inoculation) (Figures 4 and 5). Succulence is an anatomical adaption to permit the water accumulated in leaves, stems and roots to dissolve salt ions [57]. For *Elaeagnus angustifolia* L. with microbial inoculation, both the promotion of root biomass and the optimization of root morphology synthetically reduced the translation of salt ions to aerial organs, thereby mitigating the impact of toxic ions; consequently, the plants did not need to employ the succulence strategy to mediate salt stress. Meanwhile, the succulence level in root is higher than that in stem and leaf (Figures 4 and 5) and indicating that *Elaeagnus angustifolia* L. mainly relies on root succulence to increase salt tolerance [63].

Leaves, together with the stems and branches, are the carbon-fixing organs in halophytes [58,72]. Salinity stress stunts the growth of leaves, stems and branches [73], further impairing photosynthetic capacity [1,3,67,74]. Our findings provide further support for this view and showed that *Elaeagnus angustifolia* L. inoculated with AMF separately, or together with PGPR, had markedly higher Pn, Gs, Tr, and CE when compared to the control under salt stress (Figure 6), which is necessary to guarantee the photosynthesis, transpiration, CO_2 assimilation and gas exchange capacity [75]. These results are also consistent with other AMF-inoculated plant species, such as citrus [62], *Leymus chinensis* [76], and black locust [77]. The enhancement of photosynthesis may be caused by 1) growth promotion in the leaf and stem, which improves light interception and utility, and further benefits the photosynthetic capacity [28], 2) the optimization of the root structure, which is beneficial to the absorption of water and nutrients (such as nitrogen and magnesium), further promoting photosynthesis [24], and 3) combined enhancements in both gas exchange capacity and CO_2 diffusion speed are instrumental in photosynthesis [78].

The result of SFVs including 21 plant parameters showed that the salt tolerance of *Elaeagnus angustifolia* L. seedlings with AMF inoculation was higher than that in co-microbial inoculation and PGPR inoculation (Table 1). This confirmed that AMF inoculation enhances salt tolerance of *Elaeagnus angustifolia* L. seedlings, and that *Elaeagnus angustifolia* L. seedlings are more responsive to AMF inoculation than co-microbial inoculation and PGPR inoculation in saline soils. Therefore, key specific microbial species, not the microbial richness and diversity, determined the salt tolerance of one-year old *Elaeagnus angustifolia* L. seedlings in saline soil in this study [53]. Previous results showed that the plant growth-promoting effects induced by different microbial inoculations varied markedly between different growth stages and also depended on the combination of host and microbes [79]. In this study, we just measured the relevant parameters in the seedling stages of *Elaeagnus angustifolia* L. A significant hurdle remains in addressing both the longevity of the microbial inoculum in the rhizosphere and the duration of plant growth-promoting effects induced by the interactions between halophytes and microbes. This aspect needs further study in the future.

Plant growth is inseparable from leaf photosynthesis and nutrient absorption by roots. Changes in soil nutrients are further closely related to soil enzyme activities. As key indicators of soil fertility levels, soil enzyme activities play a potentially important role in biochemical reactions of soil [28]. AKP indicates the level of phosphorus in soil; UE participates in soil N cycling; SC can increase soluble nutrients in soil; CAT can prevent the toxic effects of hydrogen peroxide on plants. However, the soil enzyme activities, such as CAT, AKP, SC, and UE, are inhibited by salinity [20] due to the inhibitory effect of salt on enzyme proteins through the disruption of the tertiary protein structure [28,80]. We found that the activities of four soil enzymes all significantly increased in AMF treatments, both AKP and SC markedly increased in co-microbial inoculation, and only AKP significantly increased in PGPR treatments (Figure 7), which indicates that the effects of microbial inoculation on soil enzyme activities are enzyme-specific. Soil enzymes in the rhizosphere are most affected by AMF inoculation in terms of the measured parameters.

SFVs results show the soil enzyme activities in the AMF inoculation are higher than those in the co-microbial inoculation and PGPR inoculation, and the SFVs of soil enzyme activities is lowest in the control (Table 2). Thus, microbial inoculation can effectively promote soil enzyme activities in the rhizosphere. One possible explanation could be that the hyphal effect of AMF, the exudation of specific compounds from PGPR, and the growth promotion of roots together contribute to stimulated microbial activity and modify nutritional status in the rhizosphere under salt stress conditions [20,53,65,81]. The enhanced activities of enzymes, such as AKP and UE, could benefit the transformation of soil nutrients (such as phosphorus and nitrogen) and further further promotes the overall availability of soil nutrients. Therefore, the efforts of soil enzymes are conducive to the benign development of the soil microenvironment in this study.

5. Conclusions

Here, we observed that inoculation of *Elaeagnus angustifolia* L. with *G. mosseae* singly or synthetically with FZB42 mitigated the adverse effects of salt stress on biomass accumulation. The presence of AMF could promote root growth and modify root architecture, which may be conducive to the activities of soil enzymes and further boost nutrient acquisition. AMF inoculation also enhanced growth performance in leaves and stems, benefitting the interception and absorption of light and thus maximizing photosynthesis, transpiration, CO_2 assimilation and gas exchange capacity of *Elaeagnus angustifolia* L. in saline soil conditions. One-year-old *Elaeagnus angustifolia* L. seedlings are more responsive to the inoculation of special AMF identity per se than a combination of AMF and PGPR as well as single PGPR inoculation in saline soils. These results suggest that *Glomus mosseae* can be used in saline soil as a biotic elicits to stimulate *Elaeagnus angustifolia* L. growth and improve soil enzymes. Hence, selection of suitable microorganism inoculums is an important step before initiating any inoculation program, and could be beneficial in the recovery of saline-degraded areas.

Author Contributions: Conceptualization, J.P. and X.X.; Formal analysis, J.P. and W.Z.; Funding acquisition, C.H. and S.M.; Methodology, J.P., C.H., W.Z., and J.L.; Software, J.P. and W.Z.; Visualization, J.P., and W.Z.; Writing—original draft, J.P. and X.X.; Writing—review and editing, C.H., F.P., S.M., and X.X. All authors have read and agreed to the published version of the manuscript.

Funding: This research was funded by National Key Research and Development Program of China (No. 2017YFE0119100).

Acknowledgments: We would like to acknowledge Pro. Ruoyu Wang for his donation of the *Bacillus amyloliquefaciens* FZB42 strain and assistance in this manuscript.

Conflicts of Interest: The authors declare no conflict of interest.

References

1. Munns, R.; Tester, M. Mechanisms of salinity tolerance. *Annu. Rev. Plant Biol.* **2008**, *59*, 651–681. [CrossRef] [PubMed]

2. Estrada, B.; Aroca, R.; Azcón-Aguilar, C.; Barea, J.M.; Ruiz-Lozano, J.M. Importance of native arbuscular mycorrhizal inoculation in the halophyte *Asteriscus maritimus* for successful establishment and growth under saline conditions. *Plant Soil* **2013**, *370*, 175–185. [CrossRef]

3. Chaves, M.M.; Flexas, J.; Pinheiro, C. Photosynthesis under drought and salt stress: Regulation mechanisms from whole plant to cell. *Ann. Bot.* **2009**, *103*, 551–560. [CrossRef] [PubMed]

4. Munns, R.; Gilliham, M. Salinity tolerance of crops—What is the cost? *New Phytol.* **2015**, *208*, 668–673. [CrossRef]

5. Kosová, K.; Vítámvás, P.; Urban, M.O.; Prášil, I.T. Plant proteome responses to salinity stress–comparison of glycophytes and halophytes. *Funct. Plant Biol.* **2013**, *40*, 775–786. [CrossRef]

6. Habiba, U.; Anwarul, A.M.; Shaw, R.; Hassan, A.W.R. Salinity-Induced Livelihood Stress in Coastal Region of Bangladesh. In *Community, Environment and Disaster Risk*; Emerald Publishers: Bingley, UK, 2013.

7. Anik, A.R.; Ranjan, R.; Ranganathan, T. Estimating the impact of salinity stress on livelihood choices and incomes in Rural Bangladesh. *J. Int. Dev.* **2018**, *30*, 1414–1438. [CrossRef]

8. Rubin, R.L.; van Groenigen, K.J.; Hungate, B.A. Plant growth promoting rhizobacteria are more effective under drought: A meta-analysis. *Plant Soil* **2017**, *416*, 309–323. [CrossRef]

9. Ilangumaran, G.; Smith, D.L. Plant growth promoting rhizobacteria in amelioration of salinity stress: A systems biology perspective. *Front. Plant Sci.* **2017**, *8*, 1768. [CrossRef]

10. Chandrasekaran, M.; Boughattas, S.; Hu, S.; Oh, S.-H.; Sa, T. A meta-analysis of arbuscular mycorrhizal effects on plants grown under salt stress. *Mycorrhiza* **2014**, *24*, 611–625. [CrossRef]

11. Qian, T.; Tsunekawa, A.; Peng, F.; Masunaga, T.; Wang, T.; Li, R. Derivation of salt content in salinized soil from hyperspectral reflectance data: A case study at Minqin Oasis, Northwest China. *J. Arid Land* **2019**, *11*, 111–122. [CrossRef]

12. Huang, C.H.; Zong, L.; Buonanno, M.; Xue, X.; Wang, T.; Tedeschi, A. Impact of saline water irrigation on yield and quality of melon (*Cucumis melo cv. Huanghemi*) in northwest China. *Eur. J. Agron.* **2012**, *43*, 68–76. [CrossRef]

13. Tedeschi, A.; Zong, L.; Huang, C.H.; Vitale, L.; Volpe, M.G.; Xue, X. Effect of salinity on growth parameters, soil water potential and ion composition in *Cucumis melo cv. Huanghemi* in North-Western China. *J. Agron. Crop Sci.* **2016**, *203*, 41–55. [CrossRef]

14. Hui, H.; Tong, Q.; Feng, Z.D.; Chen, X.; Huang, A.; Youhao, E.; Gao, W. Study on salt land changes in Minqin Oasis, Northwest China. *Proc. SPIE* **2006**, *6199*. [CrossRef]

15. Qian, T.; Tsunekawa, A.; Masunaga, T.; Wang, T. Analysis of the spatial variation of soil salinity and its causal factors in China's Minqin Oasis. *Math. Probl. Eng.* **2017**, *2*, 1–9. [CrossRef]

16. Liu, Z.; Zhu, J.; Yang, X.; Wu, H.; Wei, Q.; Wei, H.; Zhang, H. Growth performance, organ-level ionic relations and organic osmoregulation of *Elaeagnus angustifolia* in response to salt stress. *PLoS ONE* **2018**, *13*, e0191552. [CrossRef] [PubMed]

17. Farzaei, M.H.; Bahramsoltani, R.; Abbasabadi, Z.; Rahimi, R. A comprehensive review on phytochemical and pharmacological aspects of *Elaeagnus angustifolia* L. *J. Pharm. Pharmacol.* **2015**, *67*, 1467–1480. [CrossRef]

18. Liu, H.; Wang, Y.; Tang, M. Arbuscular mycorrhizal fungi diversity associated with two halophytes *Lycium barbarum* L. and *Elaeagnus angustifolia* L. in Ningxia, China. *Arch. Agron. Soil Sci.* **2017**, *63*, 796–806. [CrossRef]

19. Zhang, H.; Wu, X.; Li, G.; Qin, P. Interactions between arbuscular mycorrhizal fungi and phosphate-solubilizing fungus (*Mortierella* sp.) and their effects on *Kostelelzkya virginica* growth and enzyme activities of rhizosphere and bulk soils at different salinities. *Biol. Fertil. Soils* **2011**, *47*, 543–554. [CrossRef]

20. Folli-Pereira, M.D.S.; Meira-Haddad, L.S.A.; Kasuya, M.C.M. Plant-microorganism interactions: Effects on the tolerance of plants to biotic and abiotic stresses. In *Crop Improvement: New Approaches and Modern Techniques*; Springer: Berlin, Germany, 2013; pp. 209–239.

21. Yang, X.; Yu, H.; Zhang, T.; Guo, J.; Zhang, X. Arbuscular mycorrhizal fungi improve the antioxidative response and the seed production of *Suaedoideae* species *Suaeda physophora Pall.* under salt stress. *Not. Bot. Horti Agrobot. Cluj-Napoca* **2016**, *44*, 533–540. [CrossRef]

22. Puente, E.O.R.; Prabhaharan, R.; Amador, B.M.; Espinoza, F.R.; Puente, M.; Cepeda, R.D.V. Ameliorative effects of salt resistance on physiological parameters in the halophyte *Salicornia bigelovii torr.* with plant growth-promoting rhizobacteria. *Afr. J. Biotechnol.* **2013**, *12*, 5278–5284.

23. Bharti, N.; Barnawal, D.; Maji, D.; Kalra, A. Halotolerant PGPRs prevent major shifts in indigenous microbial community structure under salinity stress. *Microb. Ecol.* **2015**, *70*, 196–208. [CrossRef] [PubMed]

24. Pan, J.; Peng, F.; Xue, X.; You, Q.; Zhang, W.; Wang, T.; Huang, C. The growth promotion of two salt-tolerant plant groups with PGPR inoculation: A meta-analysis. *Sustainability* **2019**, *11*, 378. [CrossRef]

25. Chandrasekaran, M.; Chanratana, M.; Kim, K.; Seshadri, S.; Sa, T. Impact of arbuscular mycorrhizal fungi on photosynthesis, water status, and gas exchange of plants under salt stress–A meta-analysis. *Front. Plant Sci.* **2019**, *16*, 457. [CrossRef] [PubMed]

26. Egamberdieva, D.; Wirth, S.; Jabborova, D.; Räsänen, L.A.; Liao, H. Coordination between *Bradyrhizobium* and *Pseudomonas* alleviates salt stress in soybean through altering root system architecture. *J. Plant Interact.* **2017**, *12*, 100–107. [CrossRef]

27. Ghazanfar, B.; Cheng, Z.; Cuinan, W.U.; Liu, H.; Hezi, L.I.; Rehman, R.N.U.; Ahmad, I.; Khan, A.R. *Glomus etunicatum* root inoculation and foliar application of acetyl salicylic acid induced NaCl tolerance by regulation of NaCl & LeNHXL gene expression and improved photosynthetic performance in tomato seedlings. *Pak. J. Bot.* **2016**, *48*, 1209–12117.

28. Islam, F.; Yasmeen, T.; Arif, M.S.; Ali, S.; Ali, B.; Hameed, S.; Zhou, W. Plant growth promoting bacteria confer salt tolerance in *Vigna radiata* by up-regulating antioxidant defense and biological soil fertility. *Plant Growth Regul.* **2016**, *80*, 23–36. [CrossRef]

29. Xun, F.; Xie, B.; Liu, S.; Guo, C. Effect of plant growth-promoting bacteria (PGPR) and arbuscular mycorrhizal fungi (AMF) inoculation on oats in saline-alkali soil contaminated by petroleum to enhance phytoremediation. *Environ. Sci. Pollut. Res.* **2015**, *22*, 598–608. [CrossRef]

30. Gong, Z.; Zhang, G.; Chen, Z. Development of Soil Classification in China. In *Soil Classification: A Global Desk Reference*; Eswaran, H., Ahrens, R., Rice, T.J., Stewart, B.A., Eds.; CRC Press: Boca Raton, FL, USA, 2003; pp. 101–127.

31. Soil Survey Staff, USDA. *Soil Taxonomy: A Basic System of Soil Classification for Making and Interpreting Soil Surveys*, 2nd ed.; Natural Resource Conservation Service, Agricultural Handbook No. 436; U.S. Government Printing Office: Washington, DC, USA, 1999; pp. 414–417.

32. Garg, N.; Chandel, S. The effects of salinity on nitrogen fixation and trehalose metabolism in mycorrhizal *Cajanus cajan* (L.) *Mill* sp. Plants. *J. Plant Growth Regul.* **2011**, *30*, 490–503. [CrossRef]

33. Liu, S.; Hao, H.; Lu, X.; Zhao, X.; Wang, Y.; Zhang, Y.; Xie, Z.; Wang, R. Transcriptome profiling of genes involved in induced systemic salt tolerance conferred by *Bacillus amyloliquefaciens* FZB42 in Arabidopsis thaliana. *Sci. Rep.* **2017**, *7*, 1–13. [CrossRef]

34. Hao, H.-T.; Zhao, X.; Shang, Q.-H.; Wang, Y.; Guo, Z.-H.; Zhang, Y.-B.; Xie, Z.-K.; Wang, R.-Y. Comparative digital gene expression analysis of the *Arabidopsis* response to volatiles emitted by *Bacillus amyloliquefaciens*. *PLoS ONE* **2016**, *11*, e0158621. [CrossRef]

35. Weremijewicz, J.; Janos, D.P. Arbuscular common mycorrhizal networks mediate intra- and interspecific interactions of two prairie grasses. *Mycorrhiza* **2018**, *28*, 71–83. [CrossRef] [PubMed]

36. Qi, C.-H.; Chen, M.; Song, J.; Wanga, B.S. Increase in aquaporin activity is involved in leaf succulence of the euhalophyte *Suaeda salsa* under salinity. *Plant Sci.* **2009**, *176*, 200–205. [CrossRef]

37. Han, Y.; Wang, W.; Sun, J.; Ding, M.; Zhao, R.; Deng, S.; Wang, F.; Hu, Y.; Wang, Y.; Lu, Y.; et al. *Populus euphratica* XTH overexpression enhances salinity tolerance by the development of leaf succulence in transgenic tobacco plants. *J. Exp. Bot.* **2013**, *64*, 4225–4238. [CrossRef] [PubMed]

38. Poorter, H.; Ryser, P. The limits to leaf and root plasticity: What is so special about specific root length? *New Phytol.* **2015**, *206*, 1188–1190. [CrossRef] [PubMed]

39. Giovannetti, M.; Mosse, B. An evaluation of techniques for measuring vesicular arbuscular mycorrhizal infection in roots. *New Phytol.* **1980**, *84*, 489–500. [CrossRef]

40. Sheng, M.; Tang, M.; Chen, H.; Yang, B.; Zhang, F.; Huang, Y. Influence of arbuscular mycorrhizae on the root system of maize plants under salt stress. *Rev. Can. Microbiol.* **2009**, *55*, 879–886. [CrossRef] [PubMed]

41. Luo, J.; Huang, C.; Peng, F.; Xue, X.; Wang, T. Effect of salt stress on photosynthesis and related physiological characteristics of *Lycium ruthenicum Murr*. *Acta Agric. Scand. Sect. B Soil Plant Sci.* **2017**, *67*, 680–692. [CrossRef]

42. Liu, Y.; Zhao, L.; Wang, Z.; Liu, L.; Zhang, P.; Sun, J.; Wang, B.; Song, G.; Li, X. Changes in functional gene structure and metabolic potential of the microbial community in biological soil crusts along a revegetation chronosequence in the tengger desert. *Soil Biol. Biochem.* **2018**, *126*, 40–48. [CrossRef]

43. Kandeler, E.; Tscherko, D.; Spiegel, H. Long-term monitoring of microbial biomass, N mineralization and enzyme activities of a Chernozem under different tillage management. *Biol. Fertil. Soils* **1999**, *28*, 343–351. [CrossRef]

44. Trasar-Cepeda, C.; Camiña, F.; Leirós, M.C.; Gil-Sotres, F. An improved method to measure catalase activity in soils. *Soil Biol. Biochem.* **1999**, *31*, 483–485. [CrossRef]

45. Huang, K.; Dai, X.; Xu, Y.; Dang, S.; Shi, T.; Sun, J.; Wang, K. Relation between level of autumn dormancy and salt tolerance in lucerne (*Medicago sativa*). *Crop Pasture Sci.* **2018**, *69*, 194–204. [CrossRef]

46. Munns, R. Comparative physiology of salt and water stress. *Plant Cell Environ.* **2002**, *25*, 239–250. [CrossRef] [PubMed]

47. Himabindu, Y.; Chakradhar, T.; Reddy, M.C.; Kanygin, A.; Redding, K.E.; Chandrasekhar, T. Salt-tolerant genes from halophytes are potential key players of salt tolerance in glycophytes. *Environ. Exp. Bot.* **2016**, *124*, 39–63. [CrossRef]

48. Heijden, M.G.A.V.D.; Bardgett, R.D.; Straalen, N.M.V. The unseen majority: Soil microbes as drivers of plant diversity and productivity in terrestrial ecosystems. *Ecol. Lett.* **2008**, *11*, 296–310. [CrossRef]

49. Kumar, A.; Sharma, S.; Mishra, S. Effect of alkalinity on growth performance of *Jatropha curcas* inoculated with PGPR and AM fungi. *J. Phycol.* **2009**, *3*, 177–184.

50. Hidri, R.; Barea, J.M.; Mahmoud, M.B.; Abdelly, C.; Azcón, R. Impact of microbial inoculation on biomass accumulation by *Sulla carnosa* provenances, and in regulating nutrition, physiological and antioxidant activities of this species under non-saline and saline conditions. *J. Plant Physiol.* **2016**, *201*, 28–41. [CrossRef]

51. Bharti, N.; Barnawal, D.; Shukla, S.; Tewari, S.K.; Katiyar, R.S.; Kalra, A. Integrated application of *Exiguobacterium oxidotolerans*, *Glomus fasciculatum*, and vermicompost improves growth, yield and quality of *Mentha arvensis* in salt-stressed soils. *Ind. Crops Prod.* **2016**, *83*, 717–728. [CrossRef]

52. Hemashenpagam, N.; Selvaraj, T. Effect of arbuscular mycorrhizal (AM) fungus and plant growth promoting rhizomicroorganisms (PGPR). *J. Environ. Biol.* **2011**, *32*, 579–583.

53. Vogelsang, K.M.; Reynolds, H.L.; Bever, J.D. Mycorrhizal fungal identity and richness determine the diversity and productivity of a tallgrass prairie system. *New Phytol.* **2006**, *172*, 554–562. [CrossRef]

54. Piernik, A.; Hrynkiewicz, K.; Wojciechowska, A.; Szymanska, S.; Lis, M.; Muscolo, A. Effect of halotolerant endophytic bacteria isolated from *salicornia europaea L.* on the growth of fodder beet (*Beta vulgaris L.*) under salt stress. *Arch. Agron. Soil Sci.* **2017**, *63*, 1404–1418. [CrossRef]

55. Patel, D.; Jha, C.K.; Tank, N.; Saraf, M. Growth enhancement of chickpea in saline soils using plant growth-promoting rhizobacteria. *J. Plant Growth Regul.* **2012**, *31*, 53–62. [CrossRef]

56. Niu, S.-Q.; Li, H.-R.; Paré, P.W.; Aziz, M.; Wang, S.-M.; Shi, H.; Li, J.; Han, Q.-Q.; Guo, S.-Q.; Li, J.; et al. Induced growth promotion and higher salt tolerance in the halophyte grass *Puccinellia tenuiflora* by beneficial rhizobacteria. *Plant Soil* **2016**, *407*, 217–230. [CrossRef]

57. Rewald, B.; Shelef, O.; Ephrath, J.E. Adaptive Plasticity of Salt-Stressed Root Systems. In *Ecophysiology and Responses of Plants under Salt Stress*; Springer: New York, NY, USA, 2013; pp. 169–201.

58. Poorter, H.; Niklas, K.J.; Reich, P.B.; Oleksyn, J.; Poot, P.; Mommer, L. Biomass allocation to leaves, stems and roots: Meta-analyses of interspecific variation and environmental control. *New Phytol.* **2012**, *193*, 30–50. [CrossRef] [PubMed]

59. Echeverria, M.; Scambato, A.A.; Sannazzaro, A.I.; Maiale, S.; Ruiz, O.A.; Menéndez, A.B. Phenotypic plasticity with respect to salt stress response by *Lotus glaber*: The role of its AM fungal and rhizobial symbionts. *Mycorrhiza* **2008**, *18*, 317–329. [CrossRef] [PubMed]

60. Brundrett, M. Mycorrhizas in natural Ecosystems. *Adv. Ecol. Res.* **1991**, *21*, 171–313.

61. Zou, Y.N.; Liang, Y.C.; Wu, Q.S. Mycorrhizal and non-mycorrhizal responses to salt stress in *Trifoliate Orange*: Plant growth, root architecture and soluble sugar accumulation. *Int. J. Agric. Biol.* **2013**, *15*, 565–569.

62. Wu, Q.S.; Zou, Y.N.; He, X.H. Contributions of arbuscular mycorrhizal fungi to growth, photosynthesis, root morphology and ionic balance of citrus seedlings under salt stress. *Acta Physiol. Plant.* **2010**, *32*, 297–304. [CrossRef]

63. Wang, Y.; Zhang, X.; Gao, Z.; Liu, W.; Wang, Q. Effect of salt on seedlings biomass of two varieties of *Elaeagnus* spp. *China For. Sci. Technol.* **2010**, *24*, 25–28.

64. Brundrett, M.C. Coevolution of roots and mycorrhizas of land plants. *New Phytol.* **2010**, *154*, 275–304. [CrossRef]

65. Dodd, I.C.; Perez-Alfocea, F. Microbial amelioration of crop salinity stress. *J. Exp. Bot.* **2012**, *63*, 3415–3428. [CrossRef]

66. Hodge, A. The plastic plant: Root responses to heterogeneous supplies of nutrients. *New Phytol.* **2004**, *162*, 9–24. [CrossRef]

67. Bano, A.; Fatima, M. Salt tolerance in *Zea mays* L. following inoculation with *Rhizobium* and *Pseudomonas*. *Biol. Fertil. Soils* **2009**, *45*, 405–413. [CrossRef]

68. Evelin, H.; Giri, B.; Kapoor, R. Contribution of *Glomus intraradices*, inoculation to nutrient acquisition and mitigation of ionic imbalance in NaCl-stressed Trigonella foenum-graecum. *Mycorrhiza* **2012**, *22*, 204–217. [CrossRef] [PubMed]

69. Murkute, A.A.; Sharma, S.; Singh, S.K. Studies on salt stress tolerance of citrus rootstock genotypes with arbuscular mycorrhizal fungi. *Hortic. Sci. UZPI Czech Repub.* **2006**, *33*, 70–76. [CrossRef]

70. Campanelli, A.; Ruta, C.; Mastro, G.; Morone-Fortunato, I. The role of arbuscular mycorrhizal fungi in alleviating salt stress in Medicago satival Var. Icon. *Symbiosis* **2013**, *59*, 65–76. [CrossRef]

71. Ciftci, V.; Turkmen, O.; Erdinc, C.; Sensoy, S. Effects of different arbuscular mycorrhizal fungi (AMF) species on some bean (*Phaseolus vulgaris* L.) cultivars grown in salty conditions. *Afr. J. Agric. Res.* **2010**, *5*, 3408–3416.

72. Stella, L.; Antonio, S.; Michele, P.; Teodoro, D.T.; Adriano, S. Abscisic acid root and leaf concentration in relation to biomass partitioning in salinized tomato plants. *J. Plant Physiol.* **2012**, *169*, 226–233.

73. Parida, A.K.; Das, A.B. Salt tolerance and salinity effects on plants: A review. *Ecotoxicol. Environ. Saf.* **2005**, *60*, 324–349. [CrossRef]

74. Al-Karaki, G.N.; Rusan, R.H.M. Response of two cultivars differing in salt tolerance to inoculation with mycorrhizal fungi under salt stress. *Mycorrhiza* **2001**, *11*, 43–47. [CrossRef]

75. Elhindi, K.M.; El-Din, A.S.; Elgorban, A.M. The impact of arbuscular mycorrhizal fungi in mitigating salt-induced adverse effects in sweet basil (*Ocimum basilicum* L.). *Saudi J. Biol. Sci.* **2017**, *24*, 170–179. [CrossRef]

76. Lin, J.; Wang, Y.; Sun, S.; Mu, C.; Yan, X. Effects of arbuscular mycorrhizal fungi on the growth, photosynthesis and photosynthetic pigments of *Leymus chinensis* seedlings under salt-alkali stress and nitrogen deposition. *Sci. Total Environ.* **2017**, *576*, 234–241. [CrossRef] [PubMed]

77. Zhu, X.Q.; Tang, M.; Zhang, H.Q. Arbuscular mycorrhizal fungi enhanced the growth, photosynthesis, and calorific value of black locust under salt stress. *Photosynthetica* **2017**, *55*, 378–385. [CrossRef]

78. Ahmad, P.; Azooz, M.M.; Prasad, M.N.V. *Ecophysiology and Responses of Plants under Salt Stress*; Springer: Berlin, Germany, 2013.

79. De-Bashan, L.E.; Hernandez, J.P.; Bashan, Y. The potential contribution of plant growth-promoting bacteria to reduce environmental degradation—A comprehensive evaluation. *Appl. Soil Ecol.* **2012**, *61*, 171–189. [CrossRef]

80. Rietz, D.N.; Haynes, R.J. Effects of irrigation-induced salinity and sodicity on soil microbial activity. *Soil Biol. Biochem.* **2003**, *35*, 845–854. [CrossRef]

81. Ash, C. Fungi help trees hunt for food. *Science* **2016**, *353*, 661. [CrossRef]

Article

Plant Species-Dependent Effects of Liming and Plant Residue Incorporation on Soil Bacterial Community and Activity in an Acidic Orchard Soil

Xiaodi Liu [1,2], **Zengwei Feng** [1,2], **Yang Zhou** [1,2], **Honghui Zhu** [2,*] and **Qing Yao** [1,*]

1   College of Horticulture, South China Agricultural University, Guangdong Province Key Laboratory of Microbial Signals and Disease Control, Guangdong Engineering Research Center for Litchi, Guangdong Engineering Research Center for Grass Science, Guangzhou 510642, China; liuxd@gdim.cn (X.L.); fengzw@stu.scau.edu.cn (Z.F.); zhouyang@gdim.cn (Y.Z.)

2   Guangdong Institute of Microbiology Guangdong Academy of Sciences, State Key Laboratory of Applied Microbiology Southern China, Guangdong Provincial Key Laboratory of Microbial Culture Collection and Application, Guangzhou 510070, China

*   Correspondence: zhuhh@gdim.cn (H.Z.); yaoqscau@scau.edu.cn (Q.Y.); Tel.: +86-20-87685699 (H.Z.); +86-20-85286902 (Q.Y.)

Received: 22 July 2020; Accepted: 13 August 2020; Published: 15 August 2020

Abstract: Both liming and plant residue incorporation are widely used practices for the amelioration of acidic soils—however, the difference in their effects is still not fully understood, especially regarding the microbial community. In this study, we took the acidic soils from a subtropical orchard as target soils, and implemented liming and plant residue incorporation with a leguminous and a gramineous cover crop as test plants. After six months of growth, soil pH, total organic carbon (TOC), dissolved organic carbon (DOC) and nutrient contents were determined, soil enzymes involving C, N, P cycling were assayed, and microbial communities were also analyzed using Polymerase Chain Reaction-Denaturing Gradient Gel Electrophoresis (PCR-DGGE). Results showed that liming was more effective in elevating soil pH, while plant residue incorporation exerted a more comprehensive influence—not only on soil pH, but also on soil enzyme activity and microbial community. PCR-DGGE analysis revealed that liming changed the microbial community structure more greatly than plant residue incorporation, while plant residue incorporation altered the microbial community composition much more than liming. The growth responses of test plants to liming and plant residue incorporation depended on plant species, indicating the necessity to select appropriate practice for a particular crop. A further, detailed investigation into the microbial community composition, and the respective functions using metagenomic approach, is also suggested.

Keywords: acidic soil improvement; liming; microbial community; plant residue incorporation; soil enzyme activity; subtropical orchard soil

---

1. Introduction

Acidic soils occupy around 40% of the total agricultural lands worldwide, representing one of the most important limiting factors of agriculture production [1]. In China, acidic soils mainly distribute in the subtropical regions, where hilly areas are predominant with orchards, tea gardens, forest lands as the main land-use types. Traditionally, the farmers use liming to ameliorate the soil acidity [2]. It is reported that liming increased the soil pH in a citrus orchard by 1.0~1.4 units, compared to the adjacent native forests [3]. In the tea gardens of 8-, 50-, 90-year old, liming with $CaCO_3$ at the rate of 6.4 $g·kg^{-1}$ increased the soil pH by 3.13, 3.68, 2.24 units, respectively [4]. On the other hand, these subtropical soils are typical of low organic matter content, due to fast mineralization under high temperature and humidity conditions. Consequently, plant residue incorporation is proposed and practiced as an alternative to simultaneously increase the soil pH and the soil organic matter content. Xu et al. [5] reported the increase in soil pH by up to 3.3 units after 42 days of incubation with different types of plant residues. Hue [6] demonstrated that cowpea residue incorporation not only increased soil pH from 4.45 to 5.45, but also promoted the soluble carbon from 0.11 to 2.84 mM in the Ultisol.

It is clear that liming and plant residue incorporation do not share the common mechanisms in ameliorating acidic soils. The primary effects of liming are raising pH, base saturation and Ca content, and its secondary effects include reducing Al activity and P immobilization, modifying microbial activity, and etc. (see review by Fageria and Baligar [7]). Plant residue incorporation increases soil pH as well [5,6], although the amelioration process is time-consuming compared with liming. According to Sakala et al. [8], decomposition of plant residues and the subsequent release of base and mineralized N by soil microbes is the requisite of ameliorating acidic soils by plant residue incorporation. Moreover, a large number of studies demonstrate that plant residue incorporation can directly elevate the organic matter content [9,10] and indirectly promote the microbial activity via increased soil organic carbon [11,12] and enzyme activity [13]. It seems that, apart from neutralizing soil acidity, plant residue incorporation exert more extensive influences on soil properties (e.g., microbial activity) than liming. In this context, it is reasonable to speculate that the outcomes of ameliorating soil with liming or plant residue incorporation can be much different, especially in terms of microbial properties. However, information on this aspect is scarce thus far.

Soil microbe is the essential factor driving most soil biological processes [14], including the decomposition of soil organic matter [15], nutrient (C, N, P, S) cycling [16], soil respiration [17,18], and etc. In turn, soil microbes are strongly affected by environmental cues, with pH, soil organic carbon as the important factors [19–21]. This indicates that the increase of soil pH by liming or the increase of soil organic carbon by plant residue incorporation can further regulate the microbial community. Moreover, for plant residue incorporation, the additive effect of increased pH can make the regulation complicated, which differentiates the final effects of liming and plant residue incorporation on soil microbial community. Unfortunately, the difference between them has never been investigated up to date.

We hypothesized that (1) liming would be more effective in neutralizing soil acidity than plant residues, (2) plant residue incorporation would exert a greater influence on microbial community than liming, and (3) plant residues would affect the biological properties more extensively than liming. In this study, we compared the effects of liming and plant residue incorporation in alleviating the acidity of a subtropical orchard soil. Two plant species, one legume species *Stylosanthes guianensis* and one graminoid species *Paspalum natatum*, were taken as test plants for comparison, because they are routinely grown in orchards as cover crops in the subtropical areas of China [22]. The soil chemical properties and microbial properties were monitored to evaluate the respective effects of liming and plant residue incorporation.

2. Materials and Methods

2.1. Soils and Test Plants

Soil was collected from a subtropical orchard at Heshan Hilly Land Interdisciplinary Experimental Station (E112°54′, N22°41′), Chinese Academy of Science in Guangdong province [22]. According to USDA soil classification, it is the Ultisol, with the chemical properties as follows (g·kg^{-1}): pH 4.98, total organic carbon (TOC) 15.58, dissolved organic carbon (DOC) 0.33, total N 1.64, total P 0.52, total K 10.08, available N 0.137, available P 0.067, available K 0.055. The soils were air-dried and sieved through a mesh of 2 mm pore size.

To evaluate the effects of ameliorating acidic soils by liming or plant residue incorporation, leguminous plant *S. guianensis* and gramineous plant *P. natatum* were taken as test plants. These two species are routinely grown in subtropical orchards as cover crops, and the cuttings (the aboveground biomass) are generally incorporated into orchard soil after regular mowing. The seeds of these two species used in this study were commercially obtained.

2.2. Experimental Setup

We conducted a pot culture with completely randomized factorial design in the greenhouse. Two factors were included, namely, plant species (two levels: *S. guianensis* and *P. natatum*) and soil treatments (three levels—control with no treatment, liming, and plant residue incorporation), and thus, six treatments were produced with each plant species comprising three soil treatments. For liming treatment (named as $CaCO_3$), $CaCO_3$ of chemical grade was applied to soils at the rate of 2.5 $g \cdot kg^{-1}$, equivalent to 6.5 $t \cdot ha^{-1}$. For plant residue incorporation treatment (named as OM), oven-dried, and homogenized plant shoot biomass was incorporated into the soil at the rate of 30 $g \cdot kg^{-1}$, equivalent to 78 $t \cdot ha^{-1}$. To simulate the status in the field conditions, the soil (grown with each of *S. guianensis* or *P. natatum*) was incorporated with the plant residues of respective species.

Each plastic pot (18 cm in diameter and 15 cm in height) was filled with 2.0 kg soil. Five replicates were set up for each treatment, thus, totally producing 30 pots with 15 pots for each plant species. Seeds of *S. guianensis* and *P. natatum* were surface sterilized with 10% $NaClO_2$ for 15 min, rinsed with sterilized water for five times, and then germinated at 28 °C until the primary roots of about 2 mm appeared. Approximately 50 germinated seeds were sown into each pot, and the seedlings were thinned to 20 plants of similar size per pot after two weeks of growth. Finally, 30 pots were prepared for two plant species and placed in a greenhouse with a temperature range of 23~30 °C and natural radiation. Pots were irrigated to maintain the stable water content of approximately 18% with the weighing method. To maximize the effects of ameliorating soil acidity, the soil was grown with *S. guianensis* or *P. natatum* for two growth cycles, with each lasting for three months. Briefly, after growth of three months at the first growth cycle, plants were carefully removed, and the root fragments in the soil were completely picked out, then, the germinated seeds were sown again for the second growth cycle of three months. Plant residues were incorporated into the soil only once at the beginning of the first growth cycle.

The experiment was finalized after six months of growth. Before the final harvest, the relative chlorophyll content in the leaves was measured using the portable chlorophyll meter SPAD-502 (Minolta Camera Co., Osaka, Japan). Then, plant shoots were separated from roots, and the biomass (fresh weight) of shoots and roots were recorded, respectively. The soil in each pot was homogenized and divided into three aliquots. One aliquot was stored at −80 °C for DNA extraction within two weeks, the second aliquot was stored at 4 °C for soil enzyme assay and DOC determination within three days, and the third aliquot was air-dried at room temperature for the soil chemical analysis.

2.3. Measurement of Soil Chemical Properties, TOC and DOC

Soil pH, the total and available nutrient contents (N, P, K) were determined as described previously [23]. Briefly, soil pH was measured in deionized H_2O (1:2.5 *w/v*). The contents of total N (TN), total P (TP) and total K (TK) were analyzed using the Kjeldahl method, molybdenum blue colorimetric method and flame photometer, respectively. The contents of available N (AN) was determined after the release and transformation to NH_3 by 1.07 M NaOH and $FeSO_4$ powder at 40 °C for 24 h, followed by absorption with 2% (*w/v*) H_3BO_3 and titration with 0.005 M H_2SO_4. Available P (AP) was extracted with the solution of Bray-1 (0.03 M NH_4F-0.025 M HCl), and then colorimetrically measured. Available K (AK) was extracted with 1.0 M NH_4OAc (pH = 7.0) and then determined by flame photometer.

TOC was analyzed by wet oxidation with K_2CrO_4 [24]. The extraction of DOC was according to Mavi et al. [25] with pure water as extractant, and the determination of DOC was performed using a Vario Cube TOC analyzer (Elementar Inc., Hanau, Germany).

2.4. Soil Enzyme Assay

Soil enzymes involving C, N, P cycling were assayed, including α-glucosidase (EC 3.2.1.20), β-glucosidase (EC 3.2.1.21), β-xylosidase (EC 3.2.1.37), cellobiosidase (EC 3.2.1.91), urase (EC 3.5.1.5), nitrate reductase (EC 1.7.99.4), chitinase (E.C. 3.2.1.30), acid phosphatase (EC 3.1.3.2), alkaline phosphatase (EC 3.1.3.1). The activities of α-glucosidase, β-glucosidase, β-xylosidase, cellobiosidase and chitinase was assayed using fluorogenic substrates in micro-well plates as described previously [26–28]. Briefly, soil suspensions containing 1 g fresh soil in 50 mL sodium acetate buffer (0.5 mol L^{-1}, pH 5.5) were ultrasonically homogenized, and then 160 μL aliquot of each soil suspension was dispensed into 96-well black microplates (Jet Bio-Filtration Co., Guangzhou, China). Substrate solution (final concentration of 500 μmol L^{-1}) were added to each well with a final volume of 200 μL. Negative control and quench control were also applied. Standard curves were produced as well. Microplates were incubated for 3 h at 30 °C with continuous shaking. Fluorescence was measured using a microplate fluorometer (FLx800, BioTek Instruments, Winooski, VT, USA) at 355 nm excitation and 460 nm emission. The activities of urase, nitrate reductase, acid phosphatase and alkaline phosphatase were analyzed according to Hu et al. [29] and Cui et al. [22].

2.5. Polymerase Chain Reaction-Denaturing Gradient Gel Electrophoresis Analysis of Soil Microbial Community

To compare the respective effect of CaCO$_3$ or OM on bacterial community, we performed PCR-DGGE analysis. The extraction of total soil DNA was conducted using PowerSoil® DNA isolation kit (MO BIO Laboratories Inc., Carlsbad, CA, USA) as described previously [22]. To amplify the V3 fragments of bacterial 16S rRNA genes, the nested PCR was performed using the primer sets of 27F/1492R: 5′-AGA GTT TGA TCC TGG CTC GA-3′/5′-TAC GGC TAC CTT GTT ACG ACT T-3′ (the first round [30]) and 341F-GC/518R: 5′-CGC CCG CCG CGC GCG GCG GGC GGG GCG GGG GCA CGG GGG GCC TAC GGG AGG CAG CGA-3′/5′-ATT ACC GCG GCT GCT GG-3′ (the second round [31]).

Arbuscular mycorrhizal fungi (AMF) are symbiotic soil fungi, which greatly improve host plant growth. It is reported that *P. natatum* can establish symbiosis with AMF with high colonization rate up to 99% [32]; therefore, we also probed the AMF community as affected by CaCO$_3$ or OM with PCR-DGGE analysis. To amplify the 18S rDNA fragments of AMF, the nested PCR was performed according to Wang et al. [23], with NS1/NS4: 5′-GTA GTC ATA TGC TTG TCT C-3′/5′-CTT CCG TCA ATT CCT TTA AG-3′ (the first round [33]), AML1/AML2: 5′-ATC AAC TTT CGA TGG TAG GAT AGA-3′/5′-GAA CCC AAA CAC TTT GGT TTC C-3′ (the second round [34]), NS3-GC/Glo1: 5′-CGC CCG GGG CGC GCC CCG GGC GGG GCG GGG GCA CGG GGG TTG GAG GGC AAG TCT GGT GCC-3′/5′-GCC TGC TTT AAA CAC TCTA-3′ (the third round [35]) as primer sets.

The final PCR products of bacterial community and AMF community were subjected to DGGE analysis using a D-Code Universal Mutation Detection System (Bio-Rad Laboratories, Hercules, CA, USA). For the PCR product of bacteria, DGGE were carried out on an 8% polyacrylamide gel for 10 h at a constant voltage of 80 V and 60 °C in a 45–65% horizontal denaturant gradient [31]. For the PCR product of AMF, DGGE were carried out on a 6% polyacrylamide gel for 15 h at a constant voltage of 70 V and 60 °C in a 30–60% horizontal denaturant gradient [23].

2.6. Data Analysis and Statistics

All data were the average of five replicates. Tukey's post hoc test and two-way analysis of variance (ANOVA) were performed using SPSS v21.0 (IBM SPSS In., Chicago, IL, USA). DGGE profiles were quantified with Quantity One®, and the clustering analysis was also performed with it. To evaluate the effects of pH and DOC on soil chemical properties and enzyme activities, canonical correspondence analysis (CCA) was performed using Canoco for Window 4.5 [36].

3. Results

3.1. The Effects of Liming and Plant Residue Incorporation on Plant Growth

Both $CaCO_3$ and OM affected plant growth of *S. guianensis* and *P. natatum* (Figure 1). Data indicated that OM increased, while $CaCO_3$ decreased the SPAD value of *S. guianensis* leaves. In contrast, both OM and $CaCO_3$ increased the SPAD value of *P. natatum* leaves with $CaCO_3$ more effective.

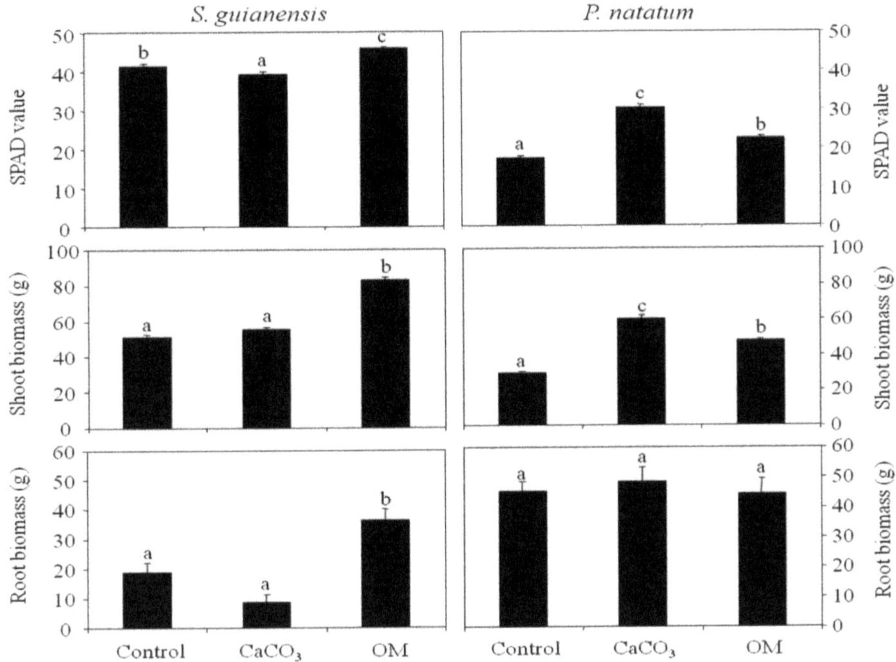

Figure 1. Influences of liming ($CaCO_3$) and plant residue incorporation (OM) on the plant growth of *S. guianensis* and *P. natatum*. Averages ($n = 5$) followed by the same letter are not significantly different (Tukey's post hoc test, $p = 0.05$).

In parallel with the SPAD value, biomass was also affected in a similar pattern (Figure 1), namely, OM increased the biomass of *S. guianensis*, while $CaCO_3$ did not affect it, and in contrast, both OM and $CaCO_3$ increased the biomass of *P. natatum* with $CaCO_3$ more effective. In detail, OM increased the biomass of *S. guianensis* by 70.4%, while $CaCO_3$ decreased by 8.6%. However, both OM and $CaCO_3$ increased the biomass of *P. natatum* by 23.4% and 44.8%, respectively. This reflects not only the difference between $CaCO_3$ and OM, but also the difference between the two plant species.

3.2. The Effects of Liming and Llant Residue Incorporation on Soil Properties

The soil pH was significantly increased by both $CaCO_3$ and OM, and plant species also exerted a significant effect on the regulation of soil pH by different treatments (Table 1). In detail, the increase in soil pH of *S. guianensis* was slightly less than that of *P. natatum* (average—0.91 vs 0.99), and the increase by $CaCO_3$ was greater than that by OM (average—1.49 vs 0.41) (Table 1).

Table 1. Influences of liming ($CaCO_3$) and plant residue incorporation (OM) on soil chemical properties. TOC, total organic carbon; DOC, dissolved organic carbon; TN, total nitrogen; AN, available nitrogen; TP, total phosphorus; AP, available phosphorus; TK, total potassium; AK, available potassium. Data followed by the same letter are not significantly different for each plant species (Tukey's post hoc, $p = 0.05$, $n = 5$).

Plant Species	Treatment	pH	TOC ($g \cdot kg^{-1}$)	DOC ($mg \cdot kg^{-1}$)	TN ($g \cdot kg^{-1}$)	AN ($mg \cdot kg^{-1}$)	TP ($g \cdot kg^{-1}$)	AP ($mg \cdot kg^{-1}$)	TK ($g \cdot kg^{-1}$)	AK ($mg \cdot kg^{-1}$)
S. guianensis	Control	4.50 ± 0.03 a	14.7 ± 0.5 a	237.1 ± 28.5 a	1.63 ± 0.03 a	121.85 ± 4.87 a	0.56 ± 0.01 a	56.25 ± 0.82 a	11.52 ± 0.43 a	17.87 ± 0.62 a
	CaCO$_3$	5.93 ± 0.03 c	16.4 ± 0.2 b	260.1 ± 8.3 a	1.64 ± 0.01 a	130.62 ± 8.15 a	0.68 ± 0.01 b	69.80 ± 0.67 b	12.28 ± 0.26 a	22.69 ± 1.19 a
	OM	4.89 ± 0.04 b	18.5 ± 0.5 c	456.8 ± 28.4 b	2.29 ± 0.02 b	135.28 ± 4.06 a	0.67 ± 0.01 b	86.45 ± 2.06 c	11.32 ± 0.21 a	58.70 ± 3.12 b
P. natatum	Control	4.89 ± 0.04 a	15.8 ± 0.2 a	251.1 ± 17.2 a	1.77 ± 0.01 b	126.92 ± 6.17 a	0.62 ± 0.01 b	61.40 ± 0.46 c	12.10 ± 0.17 b	19.06 ± 0.86 a
	CaCO$_3$	6.44 ± 0.03 c	14.8 ± 0.5 a	282.1 ± 12.0 a	1.52 ± 0.04 a	123.63 ± 5.47 a	0.55 ± 0.01 a	57.88 ± 0.87 b	11.18 ± 0.34 a	18.55 ± 0.47 a
	OM	5.31 ± 0.04 b	19.7 ± 0.6 b	423.2 ± 16.6 b	2.28 ± 0.03 c	144.46 ± 5.61 b	0.60 ± 0.011 b	47.75 ± 0.60 a	11.86 ± 0.21 a	36.79 ± 0.81 b
				Two-way ANOVA (p value)						
Plant species (P)		0.000	0.485	0.499	0.906	0.621	0.000	0.000	0.977	0.000
Soil treatments (T)		0.000	0.000	0.000	0.000	0.032	0.001	0.000	0.738	0.000
P × T		0.177	0.004	0.356	0.000	0.357	0.000	0.000	0.009	0.000

In the soil of *S. guianensis*, $CaCO_3$ significantly increased TOC, TP and AP, while OM significantly increased TOC, DOC, TN, TP, AP and AK (Table 1). AN and TK were not affected by $CaCO_3$ or OM (Table 1). Differently, in the soil of *P. natatum*, $CaCO_3$ significantly decreased TN, TP, AP and TK, while OM significantly increased TOC, DOC, TN, AN and AK, but decreased AP and TK (Table 1).

In the soil of *S. guianensis*, $CaCO_3$ significantly promoted the activities of urase, nitrate reductase and alkaline phosphatase, while OM significantly promoted the activities of β-glucosidase, β-xylosidase, urase, nitrate reductase and chitinase. The activities of α-glucosidase and cellobiosidase were not affected by either $CaCO_3$ or OM (Table 2). Differently, in the soil of *P. natatum*, $CaCO_3$ significantly promoted the activities of β-xylosidase, urase and alkaline phosphatase, while OM significantly promoted the activities of β-glucosidase, β-xylosidase, urase and nitrate reductase. The activities of α-glucosidase, cellobiosidase, chitinase and acid phosphatase were not affected by either $CaCO_3$ or OM (Table 2). Additionally, The activities of α-glucosidase, β-xylosidase, and urase were significantly higher in the soil of *S. guianensis* (average 0.24 $\mu mol \cdot L^{-1} MUF \cdot g^{-1} \cdot h^{-1}$, 0.17 $\mu mol \cdot L^{-1} MUF \cdot g^{-1} \cdot h^{-1}$, 467.6 $U \cdot g^{-1}$) than those in the soil of *P. natatum* (average 0.22 $\mu mol \cdot L^{-1} MUF \cdot g^{-1} \cdot h^{-1}$, 0.15 $\mu mol \cdot L^{-1} MUF \cdot g^{-1} \cdot h^{-1}$, 433.8 $U \cdot g^{-1}$).

3.3. The Effects of Two Practices on Soil Microbial Community

DGGE profiling showed that the bacterial community compositions of three treatments were much different from each other, and the five replicate samples of each treatment grouped well (Figure 2). Generally, the bacterial community compositions of $CaCO_3$ and Control were similar to each other, while that of OM was separated apart from $CaCO_3$ and Control for both *S. guianensis* and *P. natatum*. Moreover, the diversity index, species abundance and species evenness were not affected by either $CaCO_3$ or OM for *P. natatum* (Table 3). In contrast, for *S. guianensis*, the diversity index was significantly increased by both $CaCO_3$ and OM, while species abundance was significantly increased by only $CaCO_3$.

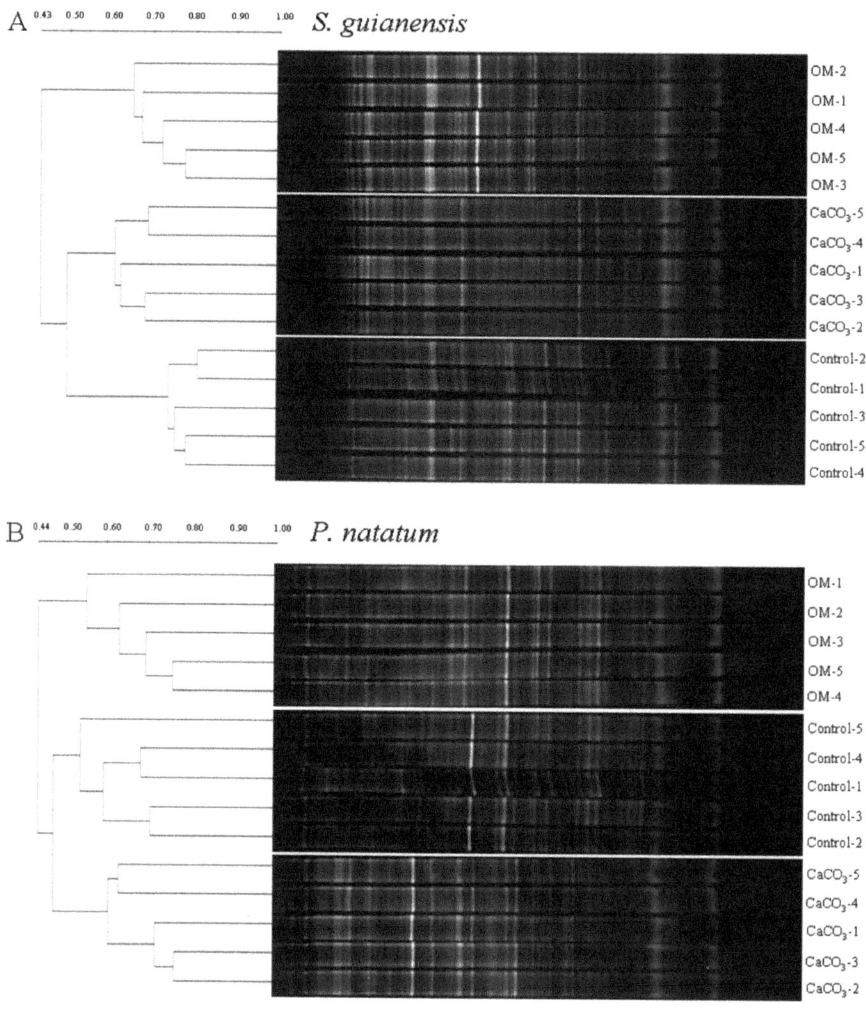

Figure 2. Denaturing Gradient Gel Electrophoresis profiles of the bacterial community in the rhizosphere of *S. guianensis* (**A**) and *P. natatum* (**B**) and the corresponding clustering analysis. CaCO$_3$, liming; OM, plant residue incorporation.

Table 2. Influences of liming ($CaCO_3$) and plant residue incorporation (OM) on soil enzyme activities. α-GLU, α-glucosidase; β-GLU, β-glucosidase; β-XYL, β-xylosidase; CEL, cellobiosidase; URA, urease; NR, nitrate reductase; CHI, chitinase; ACP, acid phosphatase; ALP, alkaline phosphatase. Data followed by the same letter are not significantly different for each plant species (Tukey's post hoc test, $p = 0.05$, $n = 5$). Enzymatic activity unit of URA, NR, ACP, ALP is $U \cdot g^{-1}$, and that of others is $\mu mol \cdot L^{-1} MUF \cdot g^{-1} \cdot h^{-1}$.

Plant Species	Treatment	α-GLU	β-GLU	β-XYL	CEL	URA	NR	CHI	ACP	ALP
S. guianensis	Control	0.23 ± 0.01 a	0.24 ± 0.01 a	0.15 ± 0.00 a	0.16 ± 0.01 a	335.4 ± 11.6 a	3.38 ± 0.07 a	0.06 ± 0.00 a	5.27 ± 0.70 ab	0.63 ± 0.11 a
	CaCO₃	0.24 ± 0.01 a	0.25 ± 0.01 a	0.16 ± 0.01 ab	0.15 ± 0.01 a	436.1 ± 9.1 b	5.09 ± 0.36 b	0.07 ± 0.00 a	7.61 ± 1.04 b	5.65 ± 0.70 b
	OM	0.25 ± 0.01 a	0.49 ± 0.04 b	0.18 ± 0.00 b	0.16 ± 0.00 a	631.3 ± 6.9 c	5.18 ± 0.22 b	0.10 ± 0.01 b	4.22 ± 1.16 a	0.73 ± 0.12 a
P. natatum	Control	0.20 ± 0.01 a	0.27 ± 0.02 a	0.13 ± 0.01 a	0.15 ± 0.00 a	197.6 ± 6.7 a	3.37 ± 0.17 a	0.08 ± 0.01 a	7.21 ± 1.15 a	0.82 ± 0.28 a
	CaCO₃	0.23 ± 0.01 a	0.33 ± 0.03 a	0.16 ± 0.01 b	0.16 ± 0.00 a	514.2 ± 16.9 b	3.55 ± 0.20 a	0.08 ± 0.01 a	4.55 ± 1.11 a	2.60 ± 0.41 b
	OM	0.23 ± 0.01 a	0.41 ± 0.02 b	0.16 ± 0.00 b	0.16 ± 0.00 a	589.7 ± 7.5 c	6.30 ± 0.11 b	0.10 ± 0.01 a	4.79 ± 0.88 a	1.71 ± 0.16 ab
					Two-way ANOVA (*p* value)					
Plant species (P)		0.036	0.497	0.004	0.800	0.001	0.420	0.150	0.820	0.052
Soil treatments (T)		0.124	0.000	0.002	0.405	0.000	0.000	0.001	0.140	0.000
P × T		0.447	0.015	0.323	0.286	0.000	0.000	0.408	0.045	0.000

Table 3. Influences of liming (CaCO$_3$) and plant residue incorporation (OM) on bacterial community structure in the soil of *S. guianensis* and *P. natatum*. Data followed by the same letter are not significantly different for each plant species (Tukey's post hoc test, $p = 0.05$, $n = 5$).

Plant Species	Treatment	Diversity Index (H)	Species Abundance (S)	Species Evenness E
S. guianensis	Control	3.64 ± 0.01 a	42.4 ± 0.4 a	0.97 ± 0.00 a
	CaCO$_3$	3.82 ± 0.01 c	49.6 ± 0.5 b	0.98 ± 0.00 a
	OM	3.69 ± 0.01 b	43.0 ± 0.5 a	0.98 ± 0.00 a
P. natatum	Control	3.76 ± 0.03 a	46.2 ± 1.1 a	0.98 ± 0.00 a
	CaCO$_3$	3.69 ± 0.05 a	45.8 ± 0.7 a	0.97 ± 0.01 a
	OM	3.79 ± 0.03 a	48.0 ± 1.1 a	0.98 ± 0.00 a
Two-way ANOVA (p value)				
Plant species (P)		0.220	0.016	0.524
Soil treatments (T)		0.119	0.001	0.210
P × T		0.000	0.000	0.101

Cluster analysis based on DGGE profiles showed that AMF community composition in the soil of *P. natatum* was also altered by both OM and CaCO$_3$, with OM more effective (Figure 3). The diversity index and species abundance of AMF community were significantly increased by CaCO$_3$ but not OM (Table 4), which is different from that of the bacterial community.

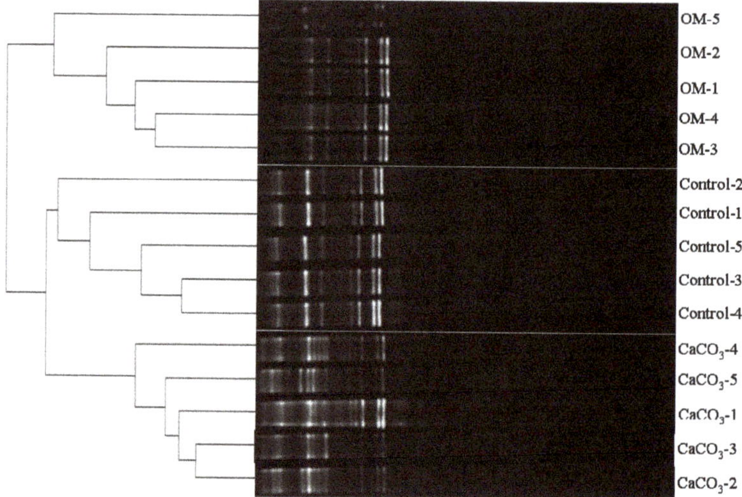

Figure 3. Denaturing Gradient Gel Electrophoresis profiles of arbuscular mycorrhizal fungal community in the soils of *P. natatum* and the corresponding clustering analysis. CaCO$_3$, liming; OM, plant residue incorporation.

Table 4. Influences of liming (CaCO$_3$) or plant residue incorporation (OM) on arbuscular mycorrhizal fungal community structure in the soil of *P. natatum*. Data followed by the same letter are not significantly different (Tukey's post hoc test, $p = 0.05$, $n = 5$).

Treatment	Diversity Index (H)	Species Abundance (S)	Species Evenness (E)
Control	2.55 ± 0.06 a	14.6 ± 0.2 a	1.06 ± 0.03 a
CaCO$_3$	2.92 ± 0.10 b	21.8 ± 0.9 b	1.05 ± 0.03 a
OM	2.54 ± 0.06 a	14.4 ± 0.2 a	1.05 ± 0.02 a

3.4. CCA of the Effects of Two Practices on Soil Properties

Since the direct effects of $CaCO_3$ and OM treatments are pH increase and DOC increase, respectively, we performed CCA to explore how $CaCO_3$ and OM affected soil chemical properties and soil enzymatic activities via pH or DOC. CCA results indicated that pH exerted little influence on soil chemical properties, while DOC showed a strong positive effect on AK and TN, but a weak negative effect on TK, AN, AP and TP (Figure 4A). For soil enzymatic activities, pH exerted a strong positive influence on ALP activity, while DOC showed a positive effect on β-glucosidase, chitinase, nitrate reductase and negative effect on acid phosphatase (Figure 4B). In combination with the data in Tables 1 and 2, these results suggest that OM extensively influenced soil chemical properties and soil enzymatic activities mainly via DOC pathway while liming influenced fewer parameters mainly via pH.

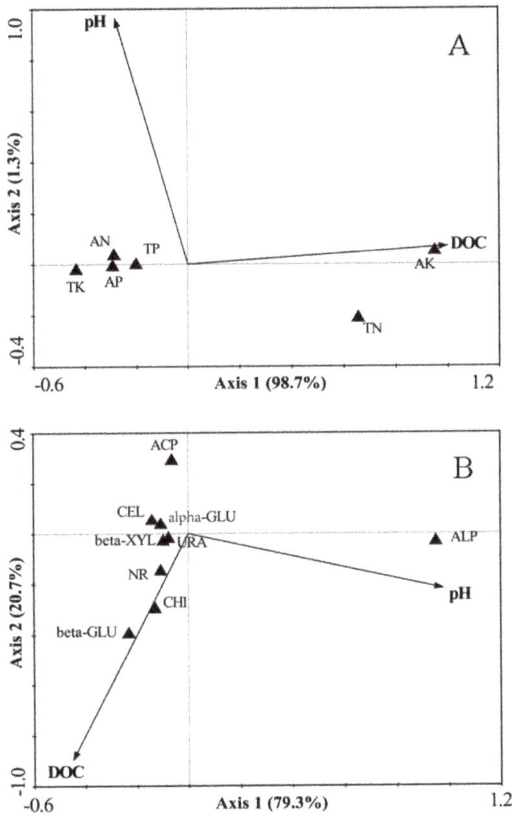

Figure 4. Canonical correspondence analysis (CCA) of the explanatory variables (pH and DOC) and the response variables, soil chemical properties (**A**) or soil enzyme activities (**B**). TN, total nitrogen; AN, available nitrogen; TP, total phosphorus; AP, available phosphorus; TK, total potassium; AK, available potassium; α-GLU, α-glucosidase; β-GLU, β-glucosidase; β-XYL, β-xylosidase; CEL, cellobiosidase; URA, urase; NR, nitrate reductase; CHI, chitinase; ACP, acid phosphatase; ALP, alkaline phosphatase.

4. Discussion

Both liming and plant residue incorporation can elevate the pH of acidic soils [3–6]; however, only a few experiments were designed to compare their effects [37,38]. In this study, we compared the effects of $CaCO_3$ and OM in increasing the pH of subtropical orchard soils with an initial pH of

4.98. It is clear that $CaCO_3$ is more effective in the amelioration of acidic soils than OM. When the effects of $CaCO_3$ and OM are compared, it should be born in mind that their primary effects are much different. $CaCO_3$ directly increases the soil pH in via soil chemical pathway, which can always be fast and obvious; however, OM indirectly increases it via soil biochemical pathway with the involvement of soil microbes. Elevating pH with plant residue incorporation involves microbial degradation of organic matter in the plant residues and the subsequent release of alkalinity (excess cations), and, thus, can be time-consuming [8]. In this scenario, the effectiveness of plant residue incorporation depends to a certain degree on the compositions and types of plant residues and characteristics of soils [5].

Although some abiotic factors, such as temperature, soil moisture, pH are the important determinants of the decomposition process of plant residues, soil microbes are the key biotic factor contributing much to the decomposition. It is generally accepted that plant residues can be decomposed by a variety of soil microbes with priming effect [39,40]. In fact, the direct effect of OM is increasing DOC due to the decomposition of plant residues [41–44], and then the promoted microbial activity. In this study, the increased pH together with promoted microbial activity, further mobilized the TN, TP, TK, leading to the increased levels of AN, AP, or AK. It is suggested that the extra-cellular hydrolysis enzymes of soil microbes are mainly responsible for the mobilization of N, P, K [45], with *Bacillus* the most effective. Our results demonstrate that the microbial community structure in limed soil is closer to that in control soil than that in the soil with plant residue incorporation, indicating a greater regulating effect on bacterial community composition by OM than that by $CaCO_3$. This difference suggests that, in parallel with the amelioration of acidic soil, OM can significantly reshape the microbial community structure while $CaCO_3$ can not. This regulated microbial community structure can further improve the microbial activity, as revealed by the soil enzyme activity in this study. This indicates that $CaCO_3$ exerted its influence on a limited array of soil chemical properties, while OM exerted its influence on a diverse array of soil chemical properties. Overall, the effects of OM in improving the soil are more comprehensive than $CaCO_3$, including not only the increased pH, but also the mobilization of nutrients and the promoted enzyme activity.

In this study, although the manifest difference in the microbial community composition was observed according to DGGE profiling, we did not characterize the dominant species or the modified species. According to Lin et al. [46], liming significantly decreased the diazotroph abundance while plant residue did not, highlighting their difference in regulating specific microbial taxa. Consequently, it is valuable to further put forward this work with 16S rRNA sequencing technique, in order that the dominant or modified species in response to $CaCO_3$ or OM can be revealed, and their respective functions can be evaluated.

In this study, we also observed the difference between plant species, including the effects on soil pH and plant growth. The soil pH in rhizosphere of *S. guianensis* was significantly lower than that of *P. natatum* regardless of treatments, mainly due to the biological N fixation (BNF) in the rhizosphere of legumes [47,48]. It is well acknowledged that the excretion of H^+ from legume roots is in parallel with BNF [49]. The differences in soil nutrient content (TP, AP, AK) and enzyme activity (α-glucosidase, β-xylosidase, urase) between plant species also existed, but showed complexity due to the interaction of plant species and soil treatments. This complexity was demonstrated previously [6,50]. Taken together, these differences led to the different response of plant growths to treatments, namely, the biomass of *S. guianensis* was greatly increased by OM, but not by $CaCO_3$, while the biomass of *P. natatum* was increased by $CaCO_3$ more greatly than by OM. The different growth responses of legumes and graminoids to liming were reported elsewhere. For example, liming at 6.72 t·ha^{-1} significantly increased the aboveground biomass of barley, but not that of peas (on the average of four years, where the initial soil pH was increased by 1.9 units) [51]. An on-farm experiment indicated that the frequency of a grain yield response to liming was almost twice for corn than for soybean [52], although the different results were also reported [53]. These results highlight the necessity of selecting the appropriate treatments (liming or plant residue incorporation) for ameliorating acidic soils when growing legumes or graminoids.

In this study, we demonstrated that the modification of the microbial community was achieved by both liming and plant residue incorporation, although the effect size of liming was weaker than that of plant residue incorporation. The modification can be the result of changes in pH and organic matter, as induced by liming and plant residue incorporation. It is well accepted that pH and organic matter are the two important factors altering microbial community composition and function [54–57]. Soil microbe is an essential biotic component in soil ecosystems, playing critical roles in diverse soil functioning. In this study, almost all soil chemical properties except AK were increased by liming or plant residue incorporation, probably due to both elevations of pH and increase in enzyme activity. The availability of most nutrients can be promoted if soil pH shifts from acidic status to neutral status with P being the most sensitive [58]. The activity of β-glucosidase decreases with soil pH increasing from 4.5 to 8.5, while organic matter increases it by providing substrates [59]. This explains well why the activity of β-glucosidase in OM treatment was higher than that in $CaCO_3$ treatment in this study. In contrast, the activity of alkaline phosphatase in $CaCO_3$ treatment was higher than that in OM treatment, most probably due to in higher pH in $CaCO_3$ treatment. Moreover, alkaline phosphatase is mainly secreted by soil microbes [60], and thus, is linked with increased TOC or DOC as demonstrated in this study. Taking together, both liming and plant residue incorporation affect soil chemical properties and biological properties, with the former more effective on chemical processes (especially pH) and the latter more effective on the biological processes. Despite the differences, both practices are able to improve acidic soil with increased nutrient availability and biological activity.

5. Conclusions

In summary, our study indicates that both liming and plant residue incorporation can ameliorate acidic soils, with liming more effective than plant residue incorporation. In parallel with pH increase, plant residue incorporation also increased the TOC and DOC contents. However, soil nutrient levels and enzyme activities were increased by liming or plant residue incorporation, the effect of plant residue incorporation was more comprehensive than that of liming. PCR-DGGE analysis revealed that liming changed the microbial community structure more greatly than plant residue incorporation, and in contrast, plant residue incorporation altered the microbial community composition much more than liming. The alteration of the microbial community composition by liming and plant residue incorporation should be characterized in detail in the future. When plant species are considered, the growth responses of *S. guianensis* (legume) and *P. natatum* (graminoid) to liming and plant residue incorporation are different, with plant residue incorporation more promotive for *S. guianensis* and liming more promotive for *P. natatum*. Our results provide a new sight in the amelioration of acidic soils, and help select appropriate practices in a combination of plants.

Author Contributions: Conceptualization, Q.Y. and H.Z.; methodology, X.L., Y.Z. and Z.F.; formal analysis, X.L. and Y.Z.; investigation, X.L., Y.Z. and Z.F.; data curation, X.L., H.Z. and Q.Y.; writing—original draft preparation, X.L. and Q.Y.; writing—review and editing, Q.Y. and H.Z.; supervision, Q.Y.; funding acquisition, Q.Y. and H.Z. All authors have read and agreed to the published version of the manuscript.

Funding: This research was funded by Guangdong Technological Innovation Strategy of Special Funds (key areas of research and development program, grant no.: 2018B020205003), and Guangdong Science and Technology Projects (2016A020210071).

Acknowledgments: We thank the colleagues at Heshan Hilly Land Interdisciplinary Experimental Station, Chinese Academy of Science for collecting the tested soil.

Conflicts of Interest: The authors declare no conflict of interest.

References

1. Sumner, M.E.; Noble, A.D. Soil acidification: The world story. In *Handbook of Soil Acidity*; Rengel, Z., Ed.; CRC Press: New York, NY, USA, 2003; pp. 1–28.

2. Crusciol, C.A.; Artigiani, A.C.; Arf, O.; Carmeis Filho, A.C.; Soratto, R.P.; Nascente, A.S.; Alvarez, R.C. Soil fertility, plant nutrition, and grain yield of upland rice affected by surface application of lime, silicate, and phosphogypsum in a tropical no-till system. *Catena* **2016**, *137*, 87–99. [CrossRef]

3. Zhu, T.; Meng, T.; Zhang, J.; Yin, Y.; Cai, Z.; Yang, W.; Zhong, W. Nitrogen mineralization, immobilization turnover, heterotrophic nitrification, and microbial groups in acid forest soils of subtropical China. *Biol. Fertil. Soils* **2013**, *49*, 323–331. [CrossRef]

4. Xue, D.; Huang, X.; Yao, H.; Huang, C. Effect of lime application on microbial community in acidic tea orchard soils in comparison with those in wasteland and forest soils. *J. Environ. Sci.* **2010**, *22*, 1253–1260. [CrossRef]

5. Xu, J.M.; Tang, C.; Chen, Z.L. The role of plant residues in pH change of acid soils differing in initial pH. *Soil Biol. Biochem.* **2006**, *38*, 709–719. [CrossRef]

6. Hue, N.V. Alleviating soil acidity with crop residues. *Soil Sci.* **2011**, *176*, 543–549. [CrossRef]

7. Fageria, N.K.; Baligar, V.C. Ameliorating soil acidity of tropical oxisols by liming for sustainable crop production. In *Advance in Agronomy*; Sparks, D.L., Ed.; Academic Press: California, SD, USA, 2008; Volume 99, pp. 345–399.

8. Sakala, G.M.; Rowell, D.L.; Pilbeam, C.J. Acid-base reactions between an acidic soil and plant residues. *Geoderma* **2004**, *123*, 219–232. [CrossRef]

9. Bierke, A.; Kaiser, K.; Guggenberger, G. Crop residue management effects on organic matter in paddy soils—The lignin component. *Geoderma* **2008**, *146*, 48–57. [CrossRef]

10. Anyanzwa, H.; Okalebo, J.R.; Othieno, C.O.; Bationo, A.; Waswa, B.S.; Kihara, J. Effects of conservation tillage, crop residue and cropping systems on changes in soil organic matter and maize-legume production: A case study in Teso District. *Nutr. Cycl. Agroecosyst.* **2010**, *88*, 39–47. [CrossRef]

11. Pascault, N.; Ranjard, L.; Kaisermann, A.; Bachar, D.; Christen, R.; Terrat, S.; Mathieu, O.; Lévêque, J.; Mougel, C.; Henault, C.; et al. Stimulation of different functional groups of bacteria by various plant residues as a driver of soil priming effect. *Ecosystems* **2013**, *16*, 810–822. [CrossRef]

12. Elmajdoub, B.; Marschner, P. Responses of soil microbial activity and biomass to salinity after repeated additions of plant residues. *Pedosphere* **2015**, *25*, 177–185. [CrossRef]

13. Geisseler, D.; Horwath, W.R.; Scow, K.M. Soil moisture and plant residue addition interact in their effect on extracellular enzyme activity. *Pedobiologia* **2011**, *54*, 71–78. [CrossRef]

14. Van der Heijden, M.G.A.; Bardgett, R.D.; van Straalen, N.M. The unseen majority: Soil microbes as drivers of plant diversity and productivity in terrestrial ecosystems. *Ecol. Lett.* **2008**, *11*, 296–310. [CrossRef] [PubMed]

15. Baumann, K.; Dignac, M.F.; Rumpel, C.; Bardoux, G.; Sarr, A.; Steffens, M.; Maron, P.A. Soil microbial diversity affects soil organic matter decomposition in a silty grassland soil. *Biogeochemistry* **2013**, *114*, 201–212. [CrossRef]

16. Broughton, R.C.I.; Newsham, K.K.; Hill, P.W.; Stott, A.; Jones, D.L. Differential acquisition of amino acid and peptide enantiomers within the soil microbial community and its implications for carbon and nitrogen cycling in soil. *Soil Biol. Biochem.* **2015**, *88*, 83–89. [CrossRef]

17. Cleveland, C.C.; Nemergut, D.R.; Schmidt, S.K.; Townsend, A.R. Increases in soil respiration following labile carbon additions linked to rapid shifts in soil microbial community composition. *Biogeochemistry* **2007**, *82*, 229–240. [CrossRef]

18. Yadav, R.L.; Shukla, S.K.; Suman, A.; Singh, P.N. Trichoderma inoculation and trash management effects on soil microbial biomass, soil respiration, nutrient uptake and yield of ratoon sugarcane under subtropical conditions. *Biol. Fertil. Soils* **2009**, *45*, 461–468. [CrossRef]

19. Rousk, J.; Bååth, E.; Brookes, P.C.; Lauber, C.L.; Lozupone, C.; Caporaso, J.G.; Knight, R.; Fierer, N. Soil bacterial and fungal communities across a pH gradient in an arable soil. *ISME J.* **2010**, *4*, 1340–1351. [CrossRef]

20. Hu, Y.; Xiang, D.; Veresoglou, S.D.; Chen, F.; Chen, Y.; Hao, Z.; Zhang, X.; Chen, B. Soil organic carbon and soil structure are driving microbial abundance and community composition across the arid and semi-arid grasslands in northern China. *Soil Biol. Biochem.* **2014**, *77*, 51–57. [CrossRef]

21. Zhalnina, K.; Dias, R.; de Quadros, P.D.; Davis-Richardson, A.; Camargo, F.A.O.; Clark, I.M.; McGrath, S.P.; Hirsch, P.R.; Triplett, E.W. Soil pH determines microbial diversity and composition in the park grass experiment. *Microb. Ecol.* **2015**, *69*, 395–406. [CrossRef]

22. Cui, H.; Zhou, Y.; Gu, Z.; Zhu, H.; Fu, S.; Yao, Q. The combined effects of cover crops and symbiotic microbes on phosphatase gene and organic phosphorus hydrolysis in subtropical orchard soils. *Soil Biol. Biochem.* **2015**, *82*, 119–126. [CrossRef]

23. Wang, C.; Gu, Z.; Cui, H.; Zhu, H.; Fu, S.; Yao, Q. Differences in arbuscular mycorrhizal fungal community composition in soils of three land use types in subtropical hilly area of southern China. *PLoS ONE* **2015**, *10*, e0130983. [CrossRef] [PubMed]

24. Deng, H.; Zhang, B.; Yin, R.; Wang, H.; Mitchell, S.M.; Griffiths, B.S.; Daniell, T.J. Long-term effect of re-vegetation on the microbial community of a severely eroded soil in sub-tropical China. *Plant Soil* **2010**, *328*, 447–458. [CrossRef]

25. Mavi, M.S.; Marschner, P.; Chittleborough, D.J.; Cox, J.W.; Sanderman, J. Salinity and sodicity affect soil respiration and dissolved organic matter dynamics differently in soils varying in texture. *Soil Biol. Biochem.* **2012**, *45*, 8–13. [CrossRef]

26. ISO/TS 22939. *Soil Quality—Measurement of Enzyme Activity Patterns in Soil Samples Using Fluorogenic Substrates in Micro-Well Plates*; International Organization for Standardization: Geneva, Switzerland, 2010.

27. Giacometti, C.; Cavani, L.; Baldoni, G.; Ciavatta, C.; Marzadori, C.; Kandeler, E. Microplate-scale fluorometric soil enzyme assays as tools to assess soil quality in a long-term agricultural field experiment. *Appl. Soil Ecol.* **2014**, *75*, 80–85. [CrossRef]

28. Karpouzas, D.G.; Kandeler, E.; Bru, D.; Friedel, I.; Auer, Y.; Kramer, S.; Vasileiadis, S.; Petric, I.; Udikovic-Kolic, N.; Djuric, S.; et al. A tiered assessment approach based on standardized methods to estimate the impact of nicosulfuron on the abundance and function of the soil microbial community. *Soil Biol. Biochem.* **2014**, *75*, 282–291. [CrossRef]

29. Hu, B.; Liang, D.; Liu, J.; Lei, L.; Yu, D. Transformation of heavy metal fractions on soil urease and nitrate reductase activities in copper and selenium co- contaminated soil. *Ecotox. Environ. Safe.* **2014**, *110*, 41–48. [CrossRef]

30. Weisburg, W.G.; Barns, S.M.; Pelletier, D.A.; Lane, D.J. 16S ribosomal DNA amplification for phylogenetic study. *J. Bacteriol.* **1991**, *173*, 697–703. [CrossRef]

31. Muyzer, G.; De Waal, E.C.; Uitterlinden, A.G. Profiling of complex microbial populations by denaturing gradient gel electrophoresis analysis of polymerase chain reaction-amplified genes coding for 16S rRNA. *Appl. Environ. Microbiol.* **1993**, *59*, 695–700. [CrossRef]

32. Ishii, T.; Matsumura, A.; Horii, S.; Motosugi, H.; Cruz, A.F. Network establishment of arbuscular mycorrhizal hyphae in the rhizospheres between citrus rootstocks and *Paspalum notatum* or *Vulpia myuros* grown in sand substrate. *Biol. Fertil. Soils* **2007**, *44*, 217. [CrossRef]

33. White, T.J.; Bruns, T.D.; Lee, S.B.; Taylor, J.W. Amplification and direct sequencing of fungal ribosomal RNA genes for phylogenetics. In *PCR Protocols and Applications—A Laboratory Manual*; Innis, N., Gelfand, D., Sninsky, J., White, T.J., Eds.; Academic: New York, NY, USA, 1990; pp. 315–322.

34. Lee, J.; Lee, S.; Young, J.P.W. Improved PCR primers for the detection and identification of arbuscular mycorrhizal fungi. *FEMS Microbiol. Ecol.* **2008**, *65*, 339–349. [CrossRef]

35. Liang, Z.; Drijber, R.A.; Lee, D.J.; Dwiekat, I.M.; Harris, S.D.; Wedin, D.A. A DGGE-cloning method to characterize arbuscular mycorrhizal community structure in soil. *Soil Biol. Biochem.* **2008**, *40*, 956–966. [CrossRef]

36. Lepš, J.; Šmilauer, P. *Multivariate Analysis of Ecological Data Using Canoco™*; Cambrige University Press: New York, NY, USA, 2003.

37. Nelson, D.R.; Mele, P.M. The impact of crop residue amendments and lime on microbial community structure and nitrogen-fixing bacteria in the wheat rhizosphere. *Aust. J. Soil Res.* **2006**, *44*, 319–329. [CrossRef]

38. Garbuio, F.J.; Jones, D.L.; Alleoni, L.R.F.; Murphy, D.V.; Caires, E.F. Carbon and nitrogen dynamics in an oxisol as affected by liming and crop residues under no-till. *Soil Sci. Soc. Am. J.* **2011**, *75*, 1723–1730. [CrossRef]

39. Kuzyakov, Y.; Friedel, J.K.; Stahr, K. Review of mechanisms and quantification of priming effects. *Soil Biol. Biochem.* **2000**, *32*, 1485–1498. [CrossRef]

40. Blagodatskaya, E.; Kuzyakov, Y. Mechanisms of real and apparent priming effects and their dependence on soil microbial biomass and community structure: Critical review. *Biol. Fert. Soils* **2008**, *45*, 115–131. [CrossRef]

41. Chantigny, M.H.; Angers, D.A.; Rochette, P. Fate of carbon and nitrogen from animal manure and crop residues in wet and cold soils. *Soil Biol. Biochem.* **2002**, *34*, 509–517. [CrossRef]

42. Chantigny, M.H. Dissolved and water-extractable organic matter in soils: A review on the influence of land use and management practices. *Geoderma* **2003**, *113*, 357–380. [CrossRef]

43. Hassan, W. C and N mineralization and dissolved organic matter potentials of two contrasting plant residues: Effects of residue type, moisture, and temperature. *Acta Agric. Scand. Sect. B-Soil Plant Sci.* **2013**, *63*, 642–652. [CrossRef]

44. Ye, G.; Lin, Y.; Liu, D.; Chen, Z.; Luo, J.; Bolan, N.; Fan, J.; Ding, W. Long-term application of manure over plant residues mitigates acidification, builds soil organic carbon and shifts prokaryotic diversity in acidic Ultisols. *Appl. Soil Ecol.* **2019**, *133*, 24–33. [CrossRef]

45. Ramesh, A.; Sharma, S.K.; Yadav, N.; Joshi, O.P. Phosphorus mobilization from native soil P-pool upon inoculation with phytate-mineralizing and phosphate-solubilizing Bacillus aryabhattai isolates for improved P-acquisition and growth of soybean and wheat crops in microcosm conditions. *Agric. Res.* **2014**, *3*, 118–127. [CrossRef]

46. Lin, Y.; Ye, G.; Liu, D.; Ledgard, S.; Luo, J.; Fan, J.; Yuan, J.; Chen, Z.; Ding, W. Long-term application of lime or pig manure rather than plant residues suppressed diazotroph abundance and diversity and altered community structure in an acidic Ultisol. *Soil Biol. Biochem.* **2018**, *123*, 218–228. [CrossRef]

47. Jarvis, S.C.; Robson, A.D. The effects of nitrogen nutrition of plants on the development of acidity in Western Australian soils. II. Effects of differences in cation/anion balance between plant species grown under non-leaching conditions. *Aust. J. Agric. Res.* **1983**, *34*, 355–365. [CrossRef]

48. Tang, C.; Barton, L.; Raphael, C. Pasture legume species differ in their capacity to acidify soil. *Aust. J. Agric. Res.* **1998**, *49*, 53–58. [CrossRef]

49. Tang, C.; McLay, C.D.A.; Barton, L. A comparison of proton excretion of twelve pasture legumes grown in nutrient solution. *Aust. J. Exp. Agric.* **1997**, *37*, 563–570. [CrossRef]

50. Vinther, F.P.; Hansen, E.M.; Olesen, J.E. Effects of plant residues on crop performance, N mineralisation and microbial activity including field CO_2 and N_2O fluxes in unfertilised crop rotations. *Nutr. Cycl. Agroecosyst.* **2004**, *70*, 189–199. [CrossRef]

51. Arshad, M.A.; Soon, Y.K.; Azooz, R.H.; Lupwayi, N.Z.; Chang, S.X. Soil and crop response to wood ash and lime application in acidic soils. *Agron. J.* **2012**, *104*, 715–721. [CrossRef]

52. Pagani, A.; Mallarino, A.P. On-farm evaluation of corn and soybean grain yield and soil pH responses to liming. *Agron. J.* **2015**, *107*, 71–82. [CrossRef]

53. Kassel, P. Report on the Effects of Eggshells and Aglime on Soil pH and Crop Yields. Northwest Research Farm and Allee Demonstration Farm Annual Report. Iowa State University, Ames. 2008. RFR—A1056. ISRF08-29, 31. Available online: http://lib.dr.iastate.edu/farms_reports/594 (accessed on 14 April 2020).

54. Judd, K.E.; Crump, B.C.; Kling, G.W. Variation in dissolved organic matter controls bacterial production and community composition. *Ecology* **2006**, *87*, 2068–2079.

55. Lauber, C.L.; Hamady, M.; Knight, R.; Fierer, N. Pyrosequencing-based assessment of soil pH as a predictor of soil bacterial community structure at the continental scale. *Appl. Environ. Microbiol.* **2009**, *75*, 5111–5120. [CrossRef]

56. Zhang, Y.; Shen, H.; He, X.; Thomas, B.W.; Lupwayi, N.Z.; Hao, X.; Thomas, M.C.; Shi, X. Fertilization shapes bacterial community structure by alteration of soil pH. *Front Microbiol.* **2017**, *8*, 1325. [CrossRef]

57. Underwood, G.J.; Michel, C.; Meisterhans, G.; Niemi, A.; Belzile, C.; Witt, M.; Dumbrell, A.J.; Koch, B.P. Organic matter from Arctic sea-ice loss alters bacterial community structure and function. *Nat. Clim. Chang.* **2019**, *9*, 170–176. [CrossRef]

58. Penn, C.J.; Camberato, J.J. A critical review on soil chemical processes that control how soil pH affects phosphorus availability to plants. *Agriculture* **2019**, *9*, 120. [CrossRef]

59. Adetunji, A.T.; Lewu, F.B.; Mulidzi, R.; Ncube, B. The biological activities of β-glucosidase, phosphatase and urease as soil quality indicators: A review. *J. Soil Sci. Plant Nutr.* **2017**, *17*, 794–807. [CrossRef]

60. Spohn, M.; Kuzyakov, Y. Distribution of microbial- and root-derived phosphatase activities in the rhizosphere depending on P availability and C allocation e coupling soil zymography with ^{14}C imaging. *Soil Biol. Biochem.* **2013**, *67*, 106–113. [CrossRef]

Review

Soil Is Still an Unknown Biological System

Paolo Nannipieri

Department of Agriculture, Food, Environment and Forestry, University of Florence, 50121 Florence, Italy; paolo.nannipieri@unifi.it

Received: 24 April 2020; Accepted: 26 May 2020; Published: 27 May 2020

Abstract: More than a thousand million cells encompassing bacteria, fungi, archaea, and protists inhabit a handful of soil. The bacterial and fungal biomass can account for 1–2 and 2–5 tha^{-1} in temperate grassland soils, respectively. Despite this huge microbial biomass, the volume occupied by microorganisms is less than 1% of the available soil volume because most micro-niches are hostile environments. Soil microorganisms and fauna play a crucial role in soil ecosystem services, and functional redundancy is a peculiar characteristic of soil as a biological system. Complex interactions are often mediated by molecular signals that occur between microbes, microbes and plants, and microbes and animals. Several microbial species have been detected in soil using molecular techniques, particularly amplicon sequencing and metagenomics. However, their activities in situ are still poorly known because the use of soil metatranscriptomics and, in particular, soil proteomics is still a technical challenge. A holistic approach with the use of labelled compounds can give quantitative information on nutrient dynamics in the soil-plant system. Despite the remarkable technical progresses and the use of imaginative approaches, there are many knowledge gaps about soil as a biological system. These gaps are discussed from a historic perspective, starting from the seven grand questions proposed by Selman A. Waksman in 1927.

Keywords: microbial interactions; rhizosphere; DNA; proteins; microbial diversity; microbial activity

1. Introduction

It is well established that soil is one of the most important natural resources, with essential functions in terrestrial ecosystems. Indeed, it supports plant growth and a myriad of living organisms; controls the fate of water in the hydrological system; and recycles the wastes and bodies of plants, animals, and microorganisms. These functions depend on the biological, chemical, and physical properties of soil. However, biological properties are more sensitive than chemical and physical properties to changes in environmental conditions, including pollution problems. This has resulted in an extensive bibliography, as shown by the several international journals and books completely dedicated to soil biological properties. However, the historic perspective is often ignored because older works, especially those created before the arrival of electric searches, are ignored. The historic perspective is important in order to have a view of the development of knowledge and thus knowledge gaps and also to promote innovative studies. Several knowledge gaps about soil as a biological system were already clear at the beginning of 1900s. In 1927, Selman Waksman proposed the following seven grand questions, which were reviewed by Mc Laren in 1977 [1] and Van Elsas and Nannipieri in 2019 [2]:

1. What organisms are active under field conditions and in what ways?
2. What associative and antagonistic influences exist among soil microflora and fauna?
3. What relationships exist between soil organic matter (SOM) transformations and soil fertility?
4. What is the meaning and significance of energy balance in soil, in particular with reference to C and N?

5. How do cultivated plants influence soil transformations?
6. How can one modify soil populations and to what ends?
7. What interrelationships exist between physicochemical conditions in soil and microbial activities?

The aim of this mini review is to discuss the advances in knowledge relative to these seven questions by discussing the main properties of soil as a biological system. I shall not refer to each question in this section of the mini review, because it can be easily done by the reader. On the contrary in the Perspective section, I shall discuss one knowledge gap still existing per each of the seven grand questions, evidencing that we have a lot still to learn about soil as a biological system. Of course, it is not possible to have an exhaustive review due to the complexity and vastness of the treated matter, which exceeds the limits of a single mini review. I apologize for the many reviews not cited, but this is a mini review dealing with a vast topic and the relative and extensive bibliography cannot be cited. In addition, I apologize for the brief discussion of the several topics of this review. For a more detailed discussion of the treated matter, I suggest that the reader consult the many cited reviews.

2. Properties of Soil as a Biological System

2.1. Physicals Structure and Distribution of Organisms in Soil

Soil is a peculiar environment for living organisms, being composed of solid, liquid, and gaseous phases. The solid phase prevails, and the extent of the liquid and gaseous phases can change depending on agricultural practices and climatic conditions. The soil structure—that is, the organization and arrangements of soil particles—influences the biological, chemical, and physical properties of soil. Soil particles differ in their size, shape, and chemical composition, and thus they can be linked with different bonds. Soil structure is a hierarchical organization because primary particles bind together to form secondary particles that can interact to form bigger particles, such as microaggregates and macroaggregates [3]. This process varies in time and space [4], and the organization of the solid particles creates differently sized pores that may be filled by water or telluric atmosphere. According to Elliott and Coleman [5], pores can explain the spatial separation of soil organisms because (i) microarthropods can only inhabit macropores; (ii) nematodes can also live in intermacroaggregate pores; (iii) protozoa, small nematodes, and fungi can also be present in intramacroaggregate or intremicroaggregate pores; and (iv) intramicroaggregate pores can only be occupied by bacteria and viruses [6]. Indeed, the pore occupancy depends on the organism size; for example, bacterial size is a few micrometres, that of fungi is less than 100 μm, and that of Acari and Collembola ranges from 100 μm to 2 mm [6]. Such separation can have important effects on soil organisms because, for example, bacteria inhabiting intramicroaggregate pores escape protozoa predation.

Soil organisms inhabit less than 1% of the available space. The microsites (hotspots) with the higher microbial abundances are those where nutrients are available, such as the rhizosphere (the soil attached to roots), the detritusphere (the soil around a plant residue), the drilosphere (soil around biopores created by earthworms), etc. [7,8]. A plausible explanation of the limited occupancy of soil volume by living organisms is the hostile environment of most soil micro-niches; conditions such as acidity; low water or O_2 availability; competition between different organisms; predation; and frequent disturbances, such as drying-rewetting and freezing-thawing cycles and the presence of toxic compounds, can inhibit the activity of soil organisms [9]. However, as shown in Table 1, the microbial biomass is huge, ranging from 1% to 5% of the organic matter of soil; these values, expressed as microbial biomass C or microbial biomass N, can range from 280 to 1940 kg·ha^{-1} and from 40 to 385 kg·ha^{-1}, respectively [10].

Table 1. Some values of microbial biomass C (MBC) and microbial biomass N (MBN) in soil (redrawn from Smith and Paul 1990, [10]).

Soil	Vegetation	MBC (kg·ha^{-1})	MBN (kg·ha^{-1})
Sandy loam	Pasture	280	40
Silt loam	Cereal-grass	288	48
Clay	Pasture	750	100
Silt	Pasture	800	309
Clay loam	Cereals	1200	240
Clay loam	Wheat	1940	385

2.2. Microbial Diversity and Microbial Functions

The microbial diversity is huge, as shown by the use of molecular techniques. According to Dini-Andreote and Van Elsas [9], "A handful of soil contains, on average, more than a thousand million cells encompassing bacteria, archaea, fungi and protists—collectively called the soil microbiome". However, molecular studies have revealed that a well-defined core set of taxa is present in soils sampled from different parts of the world [9]. Two percent of bacterial phylotypes were dominant and averaged 41% of the 16S rRNA gene sequences of surface soils from 237 locations across 18 countries of 6 continents [11]. Some of the ubiquitous and dominant phylotypes included Alphaproteobacteria (*Bradyrhizobium* spp., *Sphingomonas* sp., *Rhodoplanes* sp., *Devosia* sp., and *Kaistobacter* sp.), Betaproteobacteria (*Methylibium* sp. and *Ramlibacte*), Actinobacteria (*Streptomyce* ssp., *Salinbacterium* sp. and *Mycobacterium* sp.), Acidobacteria (*Candidatus* sp., and *Salibacter*), and Plancttomycetes. Less than 18% of the identified phylotypes matched an available genome at the 97% 16S rRNA sequence similarity level, and only 40% matched even at the 90% 16S rRNA sequence similarity level. However, the detection of a taxon does not mean it is active in soil [12]. Indeed, at least 95% of the total microbial biomass is inactive in soil, and microbial activity changes in time and space [13]. Additionally, the assumption that soil functions are those carried out by the dominant taxon under in vitro conditions may not be valid because rare taxa often play an important role in the measured microbial activity, as revealed for the alkaline phosphomonoesterase activity of soil [14] and the microbial activities of soil from a glacial retreat [15]. In addition to soil microorganisms, protozoa, nematodes, microarthropods, macroarthropods, Enchytraeidae, and Earthworms, collectively termed soil fauna, can inhabit soil [5,7]. Among the soil fauna, earthworms are the most studied and act as ecosystem engineers, with significant effects on the structure and functions of soil [16]. Viruses are also present in soil, but they have been less studied than soil organisms; they are mainly bacteriophages and can infect pathogenic bacteria as well as beneficial bacteria for plants, such as rhizobia [17]. Both the abundance and diversity of viruses are markedly affected by agricultural practices [18], and the ratio between viral and bacterial abundances is higher in soil (330-470) than in water ecosystems (1–50), but the role of viruses in biological processes and the survival of organisms in soil is poorly known [17].

Among soil organisms, bacteria and fungi play the most important role in soil processes, such as the oxidation of organic matter, including xenobiotics, N, P, S, and micronutrient transformations, and have both beneficial and negative effects on plants. [8,9]. Functional redundancy characterizes several soil metabolic processes being carried out by many different microbial species; for example, the loss of microbial diversity was thought not affect processes such as C and N mineralization [8], with a certain threshold value of microbial diversity being important [19]. However, a meta-analysis showed that the loss of diversity reduced soil C respiration by 25% [20]. It is important to underline that meta-analysis concerns different soils, plant covers, agricultural practices, and climatic conditions, whereas the comparison of the response of different microbial diversities obtained in the same soil studied under laboratory conditions avoids effects due to different parameters. However, Bao et al. [21] found that the taxonomical variability was much higher than the functional variability in bacterial communities degrading straws. Generally, processes carried out by only some microbial species, such as nitrification, are affected by a decrease in microbial diversity [8]. Microbial diversity reduced by heavy metal pollution did not affect soil organic mineralization but affected simazine degradation [22]. It is important to underline that processes such as the mineralization of specific organic compounds in soil can only be determined by using the labelled compound with the determination of the labelled carbon dioxide, because it discriminates the behavior of added C from that of soil organic C. However, by carbon dioxide there have been no insights into the sequence and type of enzymes responsible for the degradation of the organic C compound; this sequence can be different, as it occurs for the degradation of cellulose and hemicellulose [9]. Future research should study the effects of changes in microbial diversity not only on the rate of the produced carbon dioxide but also on the activities of the several enzymes which are responsible of the oxidation of the labelled compound in soil.

2.3. The Role of Important Biological Molecules Adsorbed or Entrapped in Soil

Another peculiar aspect of soil as a biological system is the ability of surface-reactive particles to adsorb important biological molecules, such as nucleic acids and proteins, and protect them from the degradation of the heterotrophic soil microbiome [8]. Enzymes released from active cells to degrade polymers to monomers or released after the death of microbial cells can be adsorbed and protected against proteolysis, remaining active [23,24]. This extracellular and stabilized enzyme activity is independent of the extant microbial activity, and it can be active under hostile conditions for microbial activity. The present enzyme assays do not distinguish the activity associated with these extracellular and stabilized enzymes from that due to microbial activity, which is due to free extracellular enzymes, enzymes attached to the outer surface of viable cells with the active site extended into the extracellular environment, and intracellular enzymes [23,24].

DNA can be released into extracellular soil environments by the lysis of dead cells [25] by the border cells of the root tip [26] and during the formation of bacterial biofilms, whose presence in soils (except for waterlogged soils) has not yet been proven [25]. DNA can move through the soil with water either by leaching or capillarity if it is not adsorbed by surface-reactive particles or degraded and used as a C, N, and P source by the heterotrophic microbial communities [25]. Adsorbed DNA can be protected against degradation by nucleases and taken up by competent bacterial cells, giving the bacterial transformation that is the incorporation of the relative gens in the genome of the host bacterial cell [25]. Despite the fact that bacterial transformation can occur at very low frequencies, it can have important implications in the gene transfer between cells located in different soil microhabitats.

2.4. Interactions between Microbes, between Microbes and Plants, and between Microbes and Fauna

Interactions can be categorized not only by the interacting species (bacteria, fungi, plant, animal or virus) but also by how cells sense neighboring other cells of the same species. This can occur at (i) the physical level, because two cells occupy the same microenvironment; (ii) the biochemical level, because a cell responds to molecular signals released from the other one; and (iii) the nutritional level, due to metabolic interactions between the two cells [27]. Of course, these three types of interactions can occur simultaneously [27]. Interactions between microbes have been extensively studied in vitro but not in soil due to the complexity and heterogeneity of soil. However, it is plausible to hypothesize that these interactions are very important in soil habitable microsites, which are isolated each from the other in dry but not in wet periods, when water connections are established between two separate microenvironments [8]. Of course, mobile organisms (such as protozoans; nematodes and earthworms), fungi, with their hyphae, chain-forming organisms, (such as *Bacillus mycoide*, actinomycetes, and fungi), and plant roots, can visit different soil microenvironments [27]. Likely, in wet periods, there is the risk of increasing anaerobic microsites due to low oxygen diffusion in water [8].

One of the most studied molecular signals between bacterial cells is the quorum sensing (QS), discovered in 1980s; it involves the regulation of several bacterial processes, such as symbiosis, virulence, competence for transformation, conjugation, antibiotic production, motility, sporulation, and biofilm formation [25]. These signals are important in the assemblage of bacterial cells, because the cells of species with specific QS signals can exclude the cells of species with anti-QS signaling traits [25]. There are also interactions not mediated by QS, like the release of antibiotics, which can either kill or inhibit the growth of the microbial partner. Bacteria can interact with fungi both positively and negatively; fungi can promote the soil colonization of bacteria inhabiting the external surface of their hyphae and using fungal exudates. Other positive interactions involve the bacterial use of monomers or oligomers produced by the extracellular breakdown of polymers, such as the cellulose of plant residues, due to the activity of extracellular enzymes released from fungi [25]. Negative interactions can also involve several mechanisms, such as the predatory action of myxobacteria on fungi, the release of antifungal compounds by bacteria, and the use of bacterial cells as nutrient sources by some fungal species [25]. These are only few examples of the several interactions occurring among microbial species in soil.

Of course, both fungi and bacteria can be used as nutrient sources by soil eukaryotic organisms, thus originating the soil food web. The microbial loop is an example of the complex interactions among bacteria, protists, and plants in nutrient dynamics [28]. Bacteria inhabiting the rhizosphere can grow due to the presence of the root exudates released from root tips; however, bacteria have to mine N from soil organic matter because root exudates are generally C-rich compounds [28]. Then, protozoa graze on bacteria, reducing their number and releasing ammonium-N because the C/N ratio of protozoa cells is higher than that of bacterial cells. The released ammonium is taken up by the plant, and thus the microbial loop shifts the competition for N between plants and bacteria in favor of plants [28]. This is just an example of the interactions between bacteria and protists, but other issues may be involved, such as selective grazing by predators [29].

The rhizosphere is the soil around the roots with a higher microbial activity than the bulk soil as the result of rhizodepostion, which includes root exudates, mucilage, root debris, and whole detached root cells [30]. Beneficial, pathogen, and neutral microorganisms inhabit the rhizosphere soil. Among the beneficial microorganisms, the plant growth-promoting bacteria can stimulate plant growth by providing nutrients and protecting plants from various abiotic and biotic stresses (among the latter, the biological control of pathogens) [30]. Some of microorganisms can positively infect plant roots, giving origin to symbiosis after a complex molecular cross-talk involving both the release of root exudates and microbial exudates. The most studied molecular cross-talks are those between rhizobia and legumes [31] and mycorrhizae and plants [32].

Generally, microbial diversity decreases when passing from the bulk to the rhizosphere soil as the result of rhizodeposition [30]. However, soil plays a fundamental role in affecting the microbial diversity of the rhizosphere. Indeed, denaturing gradient gel electrophoresis (DGGE) profiles showed that the microbial diversity of the rhizosphere soil of *Carex arenaria*, a non-mycorrhizal plant species so as to avoid confounding the effects of mycorrhizae, depended on the soil type [33]. The use of sequencing confirmed that soil could impose a larger selective pressure on plant-associated bacterial communities than plant species [34,35]. However, the soil effect decreases by comparing microbial diversity of rhizosphere soil with that of the endosphere (the interior of roots colonized by endophytes) [36]; indeed, 40% of the variation in microbial diversity depended on the host species in the endosphere but only 17% in the rhizosphere soil when 30 angiosperms were grown in a garden soil [37].

2.5. The Main Research Approaches

Soil biologists and soil biochemists have struggled and still struggle with knowing what is in soil and what it is doing, with the aim not only to better know the system but also to properly quantify soil functions [9]. They have carried out different research approaches. Studies using simplified systems, such as those only using two soil components, have brought a better understanding of interactions between soil components, such as the interactions of a single clay mineral with DNA, proteins, or microbial species [23–25,38]. Of course, caution is required to extrapolate the results of these studies on simplified systems to soil, which contains a huge variety of microbial species and, for example, not pure clay minerals, being covered by other inorganic constituents and organic material. Even DNA and proteins are released in soil with other cellular components, which may interfere with their adsorption by surface-reactive particles. This brief discussion about the drawbacks of soil omics methods and the problems with interpreting relative data does not want to discourage their use but to help with the proper discussion of the obtained data. Nowadays, omics techniques are among the most powerful ones to determine microbial diversity and microbial functions in soil.

Knowledge of the microbial taxa inhabiting soil is improved using molecular techniques, in particular amplicon sequencing and metagenome techniques. However, these studies are often based on the assumption that the detected microbial species carry out in soil the same functions observed in vitro [12]. The presence of functional genes is often taken as a sign that the relative function potentially occurs in the soil where these genes are detected without determining the gene expression or relating this presence to the measurement of the target activity [12]. Most of the microbial species inhabiting soil are generally in a dormant state, as already mentioned [13]. In addition, rare species can be active and responsible for the target activity when the expression of genes is related to the measured activity, as reported for rare species encoding alkaline phosphomonoesterase activity [14]. It is well established that metagenome studies should be combined with proteomic analyses to gain insight into soil functions because proteins are the final expression of each genome. Unfortunately, soil proteomics still has challenges and pitfalls, such as the fact that it is not technically possible to determine all expressed proteins in soil due to their huge number, and that the complete protein extraction from soil is not possible with a single extraction procedure because proteins have different properties (low and high molecular weights, hydrophilic and hydrophobic moieties, different isoelectric points) and thus different interactions with the surface-reactive particles of soil [39].

The holistic consideration of the soil system represents the best approach to quantify nutrient transformations in the soil-plant system, especially if it is combined with the use of labelled (for example, with ^{14}C or enriched with ^{15}N) compounds or the determination of changes in the abundances of stable isotopes (for example, ^{13}C) [8]. According to the holistic approach, the system is portioned into pools (for example, microbial biomass C, microbial biomass N) with a functional meaning, and fluxes between these pools can represent abiotic processes (for example, nitrate leaching or ammonia volatilization) or biotic transformations (for example, nitrification, mineralization, etc.). In this way, the distribution of the labelled compound can be monitored through the different pools and be discriminated from the respective nutrient already present in the soil (for example, the ^{15}N enriched fertilizer is distinguished by the native soil N) [8]. One limit of this approach is that the organic pool is considered as an undifferentiated whole; both N immobilization and N mineralization play a crucial role in affecting the amount of N available to plants, with the organic N pool being important in both processes with a part of it cycling more rapidly than the other part. [8]. Usually, models set up to simulate the N dynamics in soil represent both pools, but the advances in modeling N as well as C, P, and S in soil requires measurements of at least two organic pools with different biological activities.

Stable isotope probing (SIP) is a technique that combines the holistic approach with molecular methods because it can determine the active microbial population using the substrates labelled with stable isotopes (usually ^{13}C or ^{15}N); it directly links the function to the identification of species using the substrate because these species are present in heavy (labeled) DNA, which can be separated from light (unlabeled) DNA [12]. This technique can give insight not only into the stimulation of rare and dominant taxa using the labelled compound but also into the microbial taxa indirectly affected by the added compound by analyzing changes in the abundance of unlabeled taxa compared to those of the control soil (only water added) [40]. In addition, the determination of labelled phospholipid fatty acids (PLFA-SIP) can be used to trace the ^{13}C-labelled compounds released by plants pulse-labelled with ^{13}CO$_2$ not only into rhizosphere microorganisms but also into strictly plant-associated microorganisms [41]. There are two main drawbacks with the SIP technique: (i) the sensitivity of DNA-SIP is less than that of phospholipid fatty acid (PLFA)-SIP, because the former requires cell replication for incorporation and thus incubation times longer than those required for

PLFA; and (ii) the cross-feeding—that is, the labelling of microorganisms not directly involved in the substrate utilization but using the labelled metabolites produced by the users [12].

3. Perspective

Despite progression in the knowledge of the soil biological system promoted by both methodological advances and imaginative studies, many biological aspects of soil remain still obscure. Indeed, nowadays many issues about Waksman's questions are still poorly known despite more than 90 years of research. It is not possible to discuss here all these knowledge gaps due the vastness of the matter, involving the activity, abundance, and composition of soil organisms; interactions among them and with different plants; and the effects of different soil properties, environmental conditions, management practices, and polluting agents, etc. I shall only discuss one knowledge gap per each question as an example of the needed future research (Figure 1).

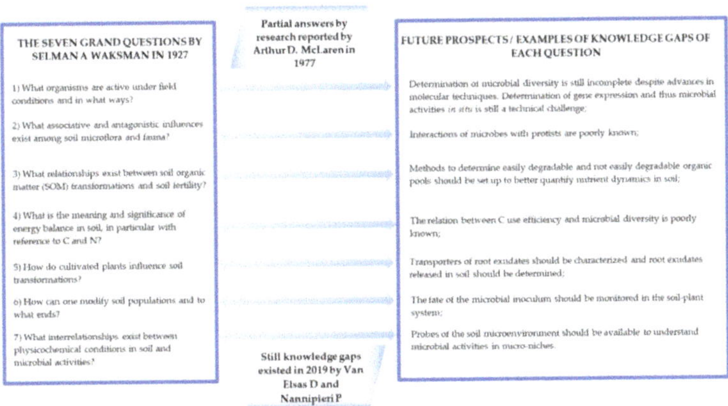

Figure 1. The seven grand questions by Selman A. Waksman and one knowledge gap for each question still present nowadays.

Concerning the first question, despite the use of molecular techniques, many species inhabiting soil are still undetected and future research should fill this gap. Several methods can determine microbial activity in soil, but the in situ determination of active microorganisms is questionable. The RNA/DNA ratio has been used as an indicator of microbial activity or microbial growth, as RNA synthesis is involved in activity and growth, whereas DNA increase is only involved in growth [42]. However, this ratio cannot be applied to clay soils, and its use does not discriminate maintenance-related synthesis versus microbial biomass production. The use of targeted transcriptomics/proteomics approaches is required but, as discussed above, soil proteomics especially are still a technical challenge.

The interactions between microorganisms and fauna are important (second question), as discussed above. However, there are several knowledge gaps about the interactions of microorganisms with all fauna components. Taking protists as an example of soil fauna, it is unknown which groups are the dominant ones in soil and, thus, which type of microbial interaction with protists is prevailing [29]

It is well established that soil organic matter plays a fundamental role in affecting soil fertility (third question). However, it is crucial to set up methods determining the easily degradable and the less degradable organic pool so as to trace main elements (C, N, P, and S) using the respective labelled compounds. This will improve the accuracy of models simulating the dynamics of these elements in soil. A great advance in soil nutrient studies was obtained by setting up the fumigation method, which enables the tracing of the dynamics of nutrients through soil microbial communities [43].

Soil microorganisms can use several organic substrates for energy provision or C supply, and this depends on several factors (fourth question), as discussed by Kästner and Miltner [44]. The C use efficiency (CUE) (the ratio between the formed microbial biomass divided by the consumed substrate) can give insight into the amount of organic C which potentially can accumulate in soil. According to Kästner and Miltner [44], it is crucial to have models predicting the organic C dynamics and energy balance in soil for the quantification of CUE and its relationship to soil microbial diversity.

It is still not possible to analyze root exudates in the presence of soil, despite their crucial role in the recruitment of microorganisms by plant roots (fifth question). Root exudates are collected under hydroponic conditions, but they differ in composition and properties when the plant is in the presence of soil. It is challenging to gain a complete extraction of root exudates from soil, not only because the extraction yield depends on the soil and exudate type but also because they are immediately used by microorganisms inhabiting the rhizosphere soil. Transporters involved in root exudation are mainly uncharacterized, and this knowledge is important in order to understand how and why plant roots release root exudates [26]. It is well established that distinct microbial communities colonize the several parts of the root due to spatially defined exudation, and the success of microbial colonization depends on chemotaxis, substrate specificity, competitiveness, and cooperativeness, but it is challenging to set up models considering all these issues.

The manipulation of the soil microbiome (sixth question) by inoculating specific microbial species in the soil-plant system for several purposes, such as controlling pathogens, improving plant growth, and bio-remediating polluted soils, has been extensively studied. However, the frequent failure of inoculated microorganisms depends on the fact that an insufficient number of inoculated active microbial cells carry out the function for which they have been inoculated in the soil-plant system. According to Gamalero and Glick [45], not only is the proper carrier of the inoculated microbial species important for the success of the inoculation, but so also is the proper understanding of the local conditions affecting the microbial inoculum. Finally, it is often a challenge to monitor the fate of the microbial inoculum in the soil-plant system, and this is crucial in order to understand the mechanisms of the effects of the inoculated microbial species [46].

Microbial activities in soil markedly depend on the soil physico-chemical properties, with temperature, moisture, and pH playing the most important role (seventh question). Several methods determine microbial activity in soil, but the most used is that determining carbon dioxide evolution from soil [8]. It is still challenging to understand how microorganisms perform their activities at the microenvironment scale. In 1982, Burns [23] concluded that it was not possible to verify experimentally the validity of his hypothesis on the ecological role of immobilized enzymes because of the lack of probes of the soil microenvironment. After almost 40 years, it is still challenging to simulate microenvironments under laboratory conditions.

4. Conclusions

This review has discussed briefly the main properties of soil as a biological system and the knowledge gaps that still persist for some of the questions raised in 1927 by Selman Waksman about the biological properties of soil. The persistence of these knowledge gaps is mainly due to the complexity of soil. Methodological progresses have allowed us to fill some of the knowledge gaps related to Waksman's questions. However, technology-driven research and hypothesis-driven research should be combined in order to fill the remaining gaps. Particularly imaginative research should address the simulation of the soil microenvironment so as to understand which factors regulate microbial activities in micro-niches. Of course, this is not an exhaustive review due to the complexity and vastness of the treated matter, which exceeds the limits of a single mini review; we suggest that the reader consult the many cited reviews to have a better view of the underlying mechanisms.

Funding: This research received no external funding.

Conflicts of Interest: The author declares no conflict of interest.

References

1. McLaren, A.D. The seven questions of Selman A. Waksman. *Soil Biol. Biochem.* **1977**, *9*, 375–376. [CrossRef]
2. Van Elsas, J.D.; Nannipieri, P. The seven grand questions on soil microbiology (Selman A. Waksman, re-examined by Arthur D. McLaren). In *Modern Soil Microbiology*, 3rd ed.; Van Elsas, J.D., Trevors, J.T., Rosado, A.S., Nannipieri, P., Eds.; CRC Press: Baca Raton, CA, USA, 2019; pp. 21–35.
3. Oades, J.M.; Waters, A.G. Aggregate hierarchy in soils. *Aust. J. Soil Res.* **1991**, *29*, 815–828. [CrossRef]
4. Tisdall, J.M.; Oades, J.M. Organic matter and water-stable aggregates in soils. *J. Soil Sci.* **1982**, *33*, 141–163. [CrossRef]
5. Elliott, E.T.; Coleman, D.C. Let the soil work for us. *Ecol. Bull. Natl. Speleol. Soc.* **1988**, *39*, 23–32.
6. Six, J. A history of research on the link between (micro) aggregates, soil biota, and soil organic matter. *Soil Tillage Res.* **2004**, *79*, 7–31. [CrossRef]
7. Coleman, D.C.; Crossley, D.A. *Fundamentals of Soil Ecology*; Academic Press: London, UK, 1996.
8. Nannipieri, P.; Ascher, J.; Ceccherini, M.T.; Landi, L.; Pietramellara, G.; Renella, G. Microbial diversity and soil functions. *Eur J. Soil Sci.* **2003**, *54*, 655–670. [CrossRef]
9. Dini-Andreote, F.; Van Elsas, J.D. The soil microbiome—An overview. In *Modern Soil Microbiology*, 3rd ed.; Van Elsas, J.D., Trevors, J.T., Rosado, A.S., Nannipieri, P., Eds.; CRC Press: Baca Raton, CA, USA, 2019; pp. 37–48.
10. Smith, J.L.; Paul, E.A. The significanceof soil microbial biomass estimations. In *Soil Biochemistry*; Bollag, J.-M., Stotzky, G., Eds.; Marcel Dekker: New York, NY, USA, 1990; Volume 6, pp. 357–396.
11. Delgado-Baquerizo, M.; Oliverio, A.M.; Brewer, T.; Benavent-Gonzalez, A.; Eldridge, D.J.; Bardgett, R.D.; Maestre, F.T.; Singh, B.K.; Fierer, N. A global atlas of the dominant bacteria found in soil. *Science* **2018**, *359*, 320–325. [CrossRef] [PubMed]
12. Nannipieri, P.; Ascher-Jenull, J.; Ceccherini, M.T.; Pietramellara, G.; Renella, G.; Schloter, M. Beyond microbial diversity for predicting soil functions: A mini review. *Pedosphere* **2020**, *30*, 5–17. [CrossRef]
13. Blagodatskaya, E.; Kuzyakov, Y. Active microorganisms in soil: Critical review of estimation criteria and approaches. *Soil Biol. Biochem.* **2013**, *67*, 192–211. [CrossRef]
14. Wei, X.; Hu, Y.; Razavi, B.S.; Zhou, J.; Shen, J.; Nannipieri, P.; Wu, J.; Ge, T. Rare taxa of alkaline phosphomonoesterase-harboring microorganisms meadiate soil phosphorus mineralization. *Soil Biol. Biochem.* **2019**, *131*, 62–70. [CrossRef]
15. Jiang, Y.; Song, H.; Lei, Y.; Korpelainen, H.; Li, C. Distinct co-occurrence patterns and driving forces of rare and abundant bacterial subcommunities following a glacial retreat in the eastern Tibetan Plateau. *Biol. Fertil. Soils* **2019**, *55*, 351–364. [CrossRef]
16. Lavelle, P.; Spain, A. *Soil Ecology*; Kluwer Academic Publisher: Dordrecht, The Netherlands, 2001.
17. Williamson, K.E.; Srinivasiah, S.; Wommack, K.E. Viruses in soil ecosystems. In *Handbook of Soil Science: Perspectives and Processes*; Huang, P.M., Li, Y., Summer, M.E., Eds.; CRC Press: Boca Raton, CA, USA, 2012; pp. 24.1–24.10.
18. Bernardo, P.C.M.; Charles-Dominique, T.; Barakat, M.; Ortet, P.; Fernandez, E.; Filloux, D.; Hartnady, P.; Rebelo, T.A.; Cousins, S.R.; Mesleard, F.; et al. Geometagenomics illuminates the impact of agriculture on the distribution and prevalence of plant viruses at the ecosystem level. *ISME J.* **2018**, *12*, 173–184. [CrossRef] [PubMed]
19. Juarez, S.; Nunan, N.; Duday, A.-N.; Pouteau, V.; Chenu, C. Soil carbon mineralisation responses to alterations of microbial diversity and soil structure. *Biol. Fertil. Soils* **2013**, *49*, 939–948. [CrossRef]
20. De Graaf, M.A.; Adkins, J.; Kardol, P.; Throop, H.L. A meta-analysis of soil biodiversity impacts on the carbon cycle. *Soil* **2015**, *1*, 257–271. [CrossRef]
21. Bao, Y.; Guo, Z.; Chen, R.; Wu, M.; Li, Z.; Lin, X.; Feng, Y. Functional community composition shows less environmental variability than 4 taxonomic composition in straw-degrading bacteria . *Biol. Fertil. Soils* **2020**, *56*. in press.
22. Singh, B.K.; Quince, C.; Macdonald, C.A.; Khachane, A.; Thomas, N.; Al-Soud, W.A.; Sorensen, S.J.; He, Z.; White, D.; Sinclair, A.; et al. Loss of microbial diversity in soils is coincident with reductions in some specialized functions. *Environ. Microbiol.* **2014**, *16*, 1–10. [CrossRef]
23. Burns, R.G. Enzyme activity in soil: Location and a possible role in microbial ecology. *Soil Biol. Biochem.* **1982**, *14*, 423–427. [CrossRef]

24. Nannipieri, P.; Trasar-Cepeda, C.; Dick, R.P. Soil enzyme activity: A brief history and biochemistry as a basis for appropriate interpretations and meta-analysis. *Biol. Fertil. Soils* **2018**, *54*, 11–19. [CrossRef]

25. Pietramellara, G.; Ascher, J.; Borgogni, F.; Ceccherini, M.T.; Guerri, G.; Nannipieri, P. Extracellular DNA in soil and sediment: Fate nd ecological relevance. *Biol. Fertil. Soils* **2009**, *45*, 219–235. [CrossRef]

26. Sasse, J.; Martinoia, E.; Nothen, T. Feed your friends: Do plant exudates shape the root microbiome? *Trends Plant. Sci.* **2018**, *23*, 25–41. [CrossRef]

27. Van Elsas, J.D.; de Araujo, W.L.; Trevors, J.T. Microbial interactions. In *Modern Soil Microbiology*, 3rd ed.; Van Elsas, J.D., Trevors, J.T., Rosado, A.S., Nannipieri, P., Eds.; CRC Press: Baca Raton, CA, USA, 2019; pp. 141–161.

28. Bonkowsky, M.; Clarholm, M. Stimulation of plant growth through interactions of bacteria and protozoa: Testing the auxiliary microbial loop hypothesis. *Acta Protozool.* **2012**, *51*, 237–247.

29. Bonkowski, M.; Dumack, K.; Fiore-Donno, A.M. The protists in soil—A token of untold eukaryotc diversity. In *Modern Soil Microbiology*, 3rd ed.; Van Elsas, J.D., Trevors, J.T., Rosado, A.S., Nannipieri, P., Eds.; CRC Press: Baca Raton, CA, USA, 2019; pp. 125–140.

30. Samad, A.; Brader, G.; Pfaffenbichler, N.; Sessitsch, A. Plant-associated bacteria and the rhizosphere. In *Modern Soil Microbiology*, 3rd ed.; Van Elsas, J.D., Trevors, J.T., Rosado, A.S., Nannipieri, P., Eds.; CRC Press: Baca Raton, CA, USA, 2019; pp. 163–178.

31. Cooper, J.E. Early interactions between legumes and rhizobia: Disclosing complexity in a molecular dialogue. *J. Appl. Microbiol.* **2007**, *103*, 1355–1365. [CrossRef] [PubMed]

32. Martin, F.M.; Perotto, S.; Bonfante, P. Mycorrhizal fungi: A fungal community at the interface between soil and roots. In *The Rhizosphere. Biochemistry and Organic Substances at the Soil-Plant Interface*, 3rd ed.; Pinton, R., Varanini, Z., Nannipieri, P., Eds.; CRC Press: Boca Raton, CA, USA, 2007; pp. 201–236.

33. De Ridder-Duine, A.S.; Kowalchuk, G.A.; Gunnewiek, P.J.A.K.; Smant, W.; van Veen, J.A.; de Boer, W. Rhizosphere bacterial community composition in natural stands of *Carex arenaria* (sand sedge) is determined by bulk soil community composition. *Soil Biol. Biochem.* **2005**, *37*, 349–357. [CrossRef]

34. Yech, Y.K.; Dennis, P.G.; Paungfoo-Lohienne, C.; Weber, L.; Brackin, R.; Ragan, M.A.; Schmidt, S.; Hugenholtz, P. Evolutionary concervation of a core root microbiome across plant phyla along a tropical soil chronosequence. *Nat. Commun.* **2017**, *8*, 215–224.

35. Schlaeppi, K.; Dombrowski, N.; Oter, R.G.; van Thernaat, E.V.L.; Schulze-Lefert, P. Quantitative divergence of the bacterial root microbiota in *Arabidopsis thaliana*. *Proc. Natl. Acad. Sci. USA* **2014**, *111*, 585–592. [CrossRef] [PubMed]

36. Escudero-Martinez, C.; Bulgarelli, D. Tracing the evolutionary routes of plant-microbiota interactions. *Curr. Opin. Microbiol.* **2019**, *49*, 34–40. [CrossRef] [PubMed]

37. Fizpatrick, C.R.; Copeland, J.; Wang, P.W.; Guttman, D.S.; Kotanen, P.M.; Johnson, M.T.J. Assembly of ecological function of the root microbiome across angiosperm plant species. *Proc. Natl. Acad. Sci. USA* **2018**, *115*, E1157–E1165. [CrossRef]

38. Stozky, G. Influence of soil mineral colloids and metabolic processes, growth adhesion, and ecology of microbes and viruses. In *Interactions of Soil Minerals with Natural Organics and Microbes*; Huang, M., Schnitzer, M., Eds.; Soil Science Society of America: Madison, WI, USA, 1986; pp. 305–428.

39. Nannipieri, P.; Giagnoni, L.; Renella, G. Metaproteomics of soil microbial communities. In *Modern Soil Microbiology*, 3rd ed.; Van Elsas, J.D., Trevors, J.T., Rosado, A.S., Nannipieri, P., Eds.; CRC Press: Baca Raton, CA, USA, 2019; pp. 257–268.

40. Bao, Y.Y.; Dolfing, J.; Wang, B.Z.; Chen, R.R.; Huang, M.S.; Li, Z.P.; Lin, X.G.; Feng, Y.Z. Bacterial communities involved directly or indirectly in the anaerobic degradation of cellulose. *Biol. Fertil. Soils* **2019**, *55*, 201–211. [CrossRef]

41. Dias, A.C.F.; Dini-Andreote, F.; Hannula, S.E.; Dini-Andreote, F.; de Cassia Pereira e Silva, M.; Salles, J.F.; de Boer, W.; van Veen, J.; van Elsas, J.D. Different selective effects on rhizosphere bacteria exerted by genetically modified versus conventional potato lines. *PLoS ONE* **2013**, *8*, e67948. [CrossRef] [PubMed]

42. Loeppmann, S.; Semenov, M.; Kuzyakov, Y.; Blagodatskaya, E. Shift from dormancy to microbial growth revealed by RNA:DNA ratio. *Soil Biol. Biochem.* **2018**, *85*, 603–612. [CrossRef]

43. Jenkinson, D.S.; Powlson, D.S. The effect of biocidal treatments on metabolism in soil. V. A method for measuring soil biomass. *Soil Biol. Biochem.* **1976**, *8*, 209–213. [CrossRef]

Appl. Sci. **2020**, *10*, 3717

44. Kästner, M.; Miltner, A. SOM and microbes—What is left from microbial life. In *The Future of Soil Carbon*; Garcia, C., Nannipieri, P., Hernandez, T., Eds.; Academic Press: London, UK, 2018; pp. 125–163.

45. Gamalero, E.; Glick, B.R. Plant growth-promoting rhizobacteria in agricultural and stressed soil. In *Modern Soil Microbiology*, 3rd ed.; Van Elsas, J.D., Trevors, J.T., Rosado, A.S., Nannipieri, P., Eds.; CRC Press: Baca Raton, CA, USA, 2019; pp. 361–380.

46. Rilling, J.I.; Acuna, J.J.; Nannipieri, P.; Cassan, F.; Maruyama, F.; Jorquera, M.A. Current opinion and perspectives on the methods for trackimg and monitoring plant growth-promoting bacteria. *Soil Biol. Biochem.* **2018**, *130*, 205–219. [CrossRef]

MDPI
St. Alban-Anlage 66
4052 Basel
Switzerland
Tel. +41 61 683 77 34
Fax +41 61 302 89 18
www.mdpi.com

Applied Sciences Editorial Office
E-mail: applsci@mdpi.com
www.mdpi.com/journal/applsci